Springer Desktop Editions in Chemistry

L. Brandsma, S.F. Vasilevsky,
H.D. Verkruijsse
Application of Transition Metal Catalysts
in Organic Synthesis
ISBN 3-540-65550-6

H. Driguez, J. Thiem (Eds.)
Glycoscience, Synthesis of Oligosaccharides
and Glycoconjugates
ISBN 3-540-65557-3

H. Driguez, J. Thiem (Eds.)
Glycoscience, Synthesis of Substrate Analogs
and Mimetics
ISBN 3-540-65546-8

K. Faber (Ed.)
Biotransformations
ISBN 3-540-66949-3

W.-D. Fessner (Ed.)
Biocatalysis, From Discovery to Application
ISBN 3-540-66970-1

S. Grabley, R. Thiericke (Eds.)
Drug Discovery from Nature
ISBN 3-540-66947-7

H.A.O. Hill, P.J. Sadler, A.J. Thomson (Eds.)
Metal Sites in Proteins and Models,
Iron Centres
ISBN 3-540- 65552-2

H.A.O. Hill, P.J. Sadler, A.J. Thomson (Eds.)
Metal Sites in Proteins and Models,
Phosphatases, Lewis Acids and Vanadium
ISBN 3-540-65553-0

H.A.O. Hill, P.J. Sadler, A.J. Thomson (Eds.)
Metal Sites in Proteins and Models,
Redox Centres
ISBN 3-540-65556-5

F.J. Leeper, J.C.Vederas (Eds.)
Biosynthesis, Polyketides and Vitamins
ISBN 3-540-66969-8

A. Manz, H. Becker (Eds.)
Microsystem Technology
in Chemistry and Life Sciences
ISBN 3-540-65555-7

P. Metz (Ed.)
Stereoselective Heterocyclic Synthesis
ISBN 3-540-65554-9

H. Pasch, B. Trathnigg
HPLC of Polymers
ISBN 3-540-65551-4

J. Rohr (Ed.)
Bioorganic Chemistry, Deoxysugars,
Polyketides and Related Classes: Synthesis,
Biosynthesis, Enzymes
ISBN 3-540-66971-X

T. Scheper (Ed.)
New Enzymes for Organic Synthesis,
Screening, Supply and Engineering
ISBN 3-540-65549-2

F.P. Schmidtchen (Ed.)
Bioorganic Chemistry,
Models and Applications
ISBN 3-540-66978-7

Springer

Berlin
Heidelberg
New York
Barcelona
Hong Kong
London
Milan
Paris
Singapore
Tokyo

F.P. Schmidtchen (Ed.)

Bioorganic Chemistry
Models and Applications

Springer

Professor Dr. Franz P. Schmidtchen
Lehrstuhl für Organische Chemie und Biochemie
Technische Universität München
Lichtenbergstr. 4
85747 München, Germany
E-mail: FPSchmid@nucleus.org.chemie.tu-muenchen.de

Description of the Series

The Springer Desktop Editions in Chemistry is a paberback series that offers selected thematic volumes from Springer chemistry series to graduate students and individual scientists in industry and academia at very affordable prices. Each volume presents an area of high current interest to a broad non-specialist audience, starting at the graduate student level.

Formerly published as hardcover edition in the review series
Topics in Current Chemistry (Vol. 184) ISBN 3-540-61388-9

Cataloging-in-Publication Data applied for

ISBN 3-540-66978-7
Springer-Verlag Berlin Heidelberg New York

Die Deutsche Bibliothek - CIP-Einheitsaufnahme
Bioorganic chemistry: models and applications / F.P. Schmidtchen (ed.). - Berlin Heidelberg, New
York; Barcelona; Hong Kong; London; Milan; Paris; Singapore; Tokyo: Springer, 2000
(Springer desktop editions in chemistry)
ISBN 3-540-66978-7

Springer-Verlag is a company in the specialist publishing group BertelsmannSpringer
© Springer-Verlag Berlin Heidelberg 2000

Cover: design & production, Heidelberg
Typesetting: Fotosatz-Service Köhler OHG, Würzburg
Printed on acid-free paper   SPIN: 10720848   02/3020 hu - 5 4 3 2 1 0

Preface

Bioorganic Chemistry more and more extends beyond the mere study of biological phenomena by the methods and paradigms of Organic Chemistry. The increasing knowledge and comprehension of biological function principles and structures inspires us to attempt to make use of the novel insights in various artifical applications.

One area of particularly promise is the engineering of biopolymers in order to make them fold into a predefined 3 dimensional structure. This is a prerequisite to the construction of nanometer sized units with fixed geometry that can sustain a desired function and eventually may lead into "nanotechnology" which is the prudent exploitation of dedicated noncovalent interactions that have been unraveled before with small model compounds.

Just as fascinating as the adaption of architectural concepts is the mimicking of biological reactivity. Quite a number of important reactions in biology do not have an equivalent in organic chemistry. Among the most amazing reactions is the selective oxidation of non-activated C-H bonds by Cytochrome P_{450} enzymes. Taking these biocatalysts as shining examples, one can hope to learn how to generate and handle highly reactive oxygen species while still preserving selectivity. The enzyme models are built according to the current understanding of the important factors involved in the biochemical mechanism. This approach to artificial enzyme-like catalysts is powered by the vision that the prominent features of enzymes in terms of selectivity, reactivity and tunable regulation may be shown by non-proteinogenic, stable molecules, too, if only the enzymatic working principles can be adapted correctly.

The desire for a sustainable development in chemistry lays the foundation for environmentally benign processes. From the view point of organic chemistry, the construction of carbon skeletons plays the pivotal rôle. The extraordinarily mild reaction conditions in addition to the non-toxic and non-burnable properties and ubiquitous availability of water as the reaction medium make enzyme-catalyzed C-C-bond formation the first choice even for industrial production. Thanks to subtle selectivity features of the corresponding enzymes a rather broad range in substrate specificity meets with a highly conserved stereospecificity at the newly connected carbon centers. In addition, these features and the availability of the respective biocatalysts are open to intervention by recombinant genetechnological techniques.

The present volume combines reports on the current status in three extensively investigated fields of bioorganic applications which are being continuously fed by the progress achieved in understanding the fundamental natural phenomena. This book is also meant to inspire the reader to look out for improvements in the approaches described here, so that we will eventually witness the substantiation of what at the moment is seen only as a bright perspective.

München, May 1996   F. P. Schmidtchen

Table of Contents

The Development of Peptide Nanostructures

Normand Voyer

Département de chimie, Université de Sherbrooke, Sherbrooke, Québec, Canada J1K 2R1

Table of Contents

List of Abbreviations

For simplicity, the one letter code is used for the amino acids: A, Alanine; C, cysteine; D, aspartic acid; E, glutamic acid, F, phenylalanine; G, glycine; H, histidine; I, isoleucine; K, lysine; L, leucine; M, methionine; N, asparagine; P, proline; Q, glutamine; R, arginine; S, serine; T, threonine; V, valine; W, tryptophane; Y, tyrosine. The D-amino acids are noted by the (D) symbol preceding the one letter code. CD, circular dichroism spectropolarimetry.

Topics in Current Chemistry, Vol. 184
© Springer Verlag Berlin Heidelberg 1996

This chapter summarizes the recent developments in the preparation of structurally defined peptidic molecules on the nanometer scale. Emphasis is placed on the unnatural strategies to stabilize peptidic conformations and on the recent examples of functional peptide-based molecular devices rather than peptide synthesis itself.

1 Introduction

The development of nanometer-scale molecules that are structurally well-defined is currently of great interest. Not only do these molecules have important applications in material science, molecular electronics, and bioorganic chemistry, but their synthesis and characterization also represent a great challenge for chemists. As in many areas of research, Nature herself, with an infinite number of functional nanostructures, is a source of inspiration. Indeed, in the course of evolution, Nature has developed enzymes and proteins of nanometer size that perform catalysis and molecular recognition in an exquisite fashion. And it is by no means accidental that polypeptides were selected as the construction materials of these sophisticated molecular systems. First, the amino acids with their bifunctional (nucleo- and electrophilic) character, their chiral nature, and their 20 different side chains are ideal monomers for the construction of linear polymers. Secondly, the peptide bond is stable to hydrolysis and brings rigidity to the polypeptide chains. It also forces the chains to fold into specific conformations in solution, i.e. secondary structures such as the α-helix, the β-sheet, and the turns that are the constituents of the globular proteins. Therefore, it is not surprising that special efforts have been devoted to the development of peptide nanostructures in recent years [1]. The synthetic approach of chemists is complementary to the molecular-biology approach (directed mutagenesis, molecular biology) and aims at developing artificial molecular devices which will be as efficient as the natural molecular machinery. This chapter summarizes the recent progress made by chemists in the development of peptide nanostructures.

2 Structurally-Defined Peptide Nanostructures

The preparation and characterization of short peptidic molecules that adopt a stable and predictible structure in solution is a prerequisite for the construction of *de novo*-designed artificial enzymes and proteins. In natural polypeptides, the secondary structures are parts of a larger system and their conformational stability is due to several intra- and interchain non-covalent interactions such as van der Waals' forces, electrostatic forces, hydrogen bonding, and hydrophobic forces [2]. However, these interactions are less important in short (up to 15 residues) monomeric peptides and their conformational equilibrium in water usually lies towards the more entropically favored structureless form rather than the ordered form (Fig. 1) [3].

Two strategies have been used to overcome this difficulty and to help prepare peptide segments with a well defined three dimensional structure.

2.1 Using Unnatural Side Chain Interactions

One of the strategies is to "engineer" in a peptide structure two distant unnatural side chains that can interact either directly or through the selective complexation of a difunctional guest in between the two side chains. Depending on the distance between the two residues bearing the binding sites, the engineered interactions can stabilize a specific conformation which will bring the binding elements into the required orientation for the interactions.

Baldwin and co-workers [4] have made a pioneering contribution in this area. They demonstrated that short (11–15 residues) alanine-rich peptides (such as 1), having several glu-lys salt bridges adopt a stable monomeric α-helix structure in solution.

Ac-A-E-A$_3$-K-E-A$_3$-K-E-A$_3$-K-A-NH$_2$ **1**

When the side chain interactions proceed through recognition of a specific guest, they have the additional feature that specific conformations can be induced and modulated by the reversible complexation of an effector molecule. Also, the conformation induced in this latter process is the one that brings the binding sites in a complementary fashion to the guest geometry (Fig. 2). Therefore, the complexation of guests with differing geometries can induce, in principle, various secondary structures.

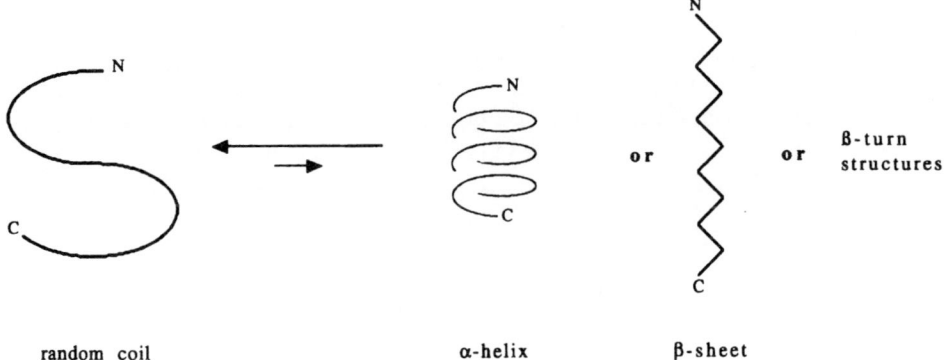

random coil   α-helix   β-sheet

Fig. 1. The conformational equilibrium in short peptidic structures lies towards the structureless form

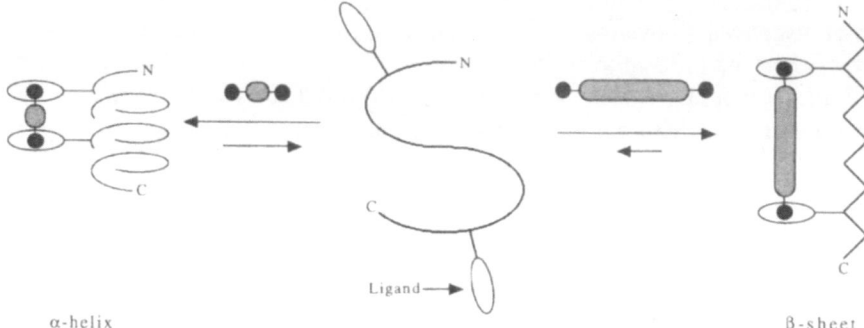

Fig. 2. Schematic representation of the use of unnatural side chain interactions to alter the conformational equilibrium towards a more ordered structure. The binding of a specific guest displaces the equilibrium towards a conformation that orients the binding side chains complementary to the geometry of the guest

This strategy is somewhat reminescent of the biological folding pathways of natural polypeptides that use selective interactions with effectors (Ca^{2+}, chaperone proteins, etc.) to start the nucleation of the secondary structures [5] and has been used successfully to prepare well defined peptide nanostructures by several groups.

Hopkins and co-workers [6] have used the selective complexation of transition metals by two distant EDTA modified amino acids to stabilize the α-helical conformation of peptides **2** and **3** (Fig. 3). The results were particularly impressive in the case of **3** where the helicity increased from 0 to about 80% upon complexation of Cd^{2+} ions. Along the same lines, Ghadiri and coworkers [7] reported the important stabilization of the helical conformation of **4** and **5** by the formation of selective metal complexes (Ru^{2+}, Zn^{2+}, Cu^{2+}, and Cd^{2+}) involving either two imidazoles of histidines or one imidazole and one thiol from a cysteine separated by three amino acids (I, I + 4) (Fig. 4). They also reported that peptide **4** is Cd^{2+}-selective and that the helical conformation of the inert Ru^{2+} complex of **5** is remarkably stable. For instance, it has a melting point 25 °C higher than the uncomplexed peptide in water.

Another interesting approach to solve the problem of preparing peptide nanostructures with predictable solution conformations has been taken by Gellman and Dado [8]. They designed an 18-residue peptide **6** that could have

$$\text{Ac-NH-}\underset{\overset{\displaystyle |}{\text{H}}}{\overset{\overset{\displaystyle \text{R}}{|}}{\text{C}}}\text{-CO-A}_3\text{-NH-}\underset{\overset{\displaystyle |}{\text{H}}}{\overset{\overset{\displaystyle \text{R}}{|}}{\text{C}}}\text{-CO-(A}_4\text{-E-K)}_3\text{-NH}_2 \quad \textbf{2}$$

$$\text{Ac-NH-}\underset{\overset{\displaystyle |}{\text{H}}}{\overset{\overset{\displaystyle \text{R}}{|}}{\text{C}}}\text{-CO-A}_3\text{-NH-}\underset{\overset{\displaystyle |}{\text{H}}}{\overset{\overset{\displaystyle \text{R}}{|}}{\text{C}}}\text{-CO-A}_4\text{-E-K}_3\text{-NH}_2 \quad \textbf{3}$$

$R = -N(CH_2COOH)_2$

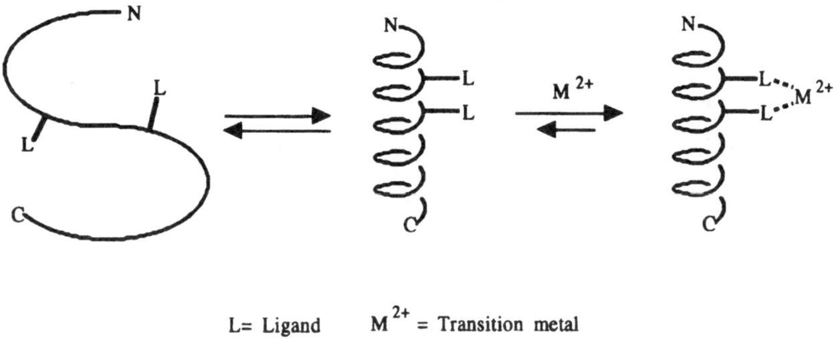

L= Ligand M^{2+} = Transition metal

Fig. 3. The stabilization of the helical form of a peptide by the selective formation of a complex between two aminodiacetic-acid modified side chains and a transition metal. (Reproduced with the permission of Ref. 1)

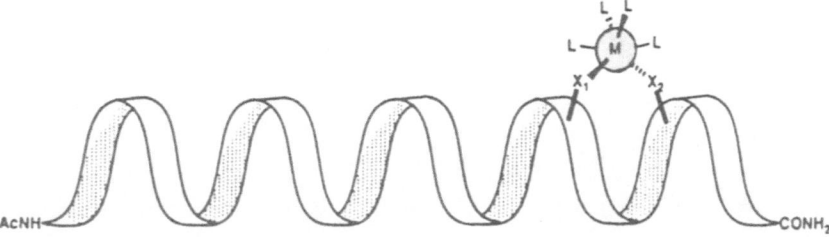

Fig. 4. Illustration of an α-helical peptide stabilized by a transition metal complex involving two histidine or one histidine and one cysteine side chains. (Reproduced with the permission of Ref. 7)

Ac-A-E-A$_3$-K-E-A$_3$-K-H-A$_3$-H-A-NH$_2$ **4**

Ac-A-E-A$_3$-K-E-A$_3$-K-C-A$_3$-H-A-NH$_2$ **5**

Ac-Y-L-K-A-M-K-E-A-M-A-K-L-M-A-K-L-M-A-NH$_2$ **6**

its secondary structure modulated between a β-sheet and an α-helix by a simple redox process. The peptide is composed of hydrophobic amino acids (leucines, alanines), hydrophilic amino acids (lysines, glutamic acid), and four methionines. In their reduced form the latter are essentially hydrophobic and **6** adopts a stable α-helical conformation creating an amphiphilic structure (Fig. 5). However, upon oxidation to the sulfoxide form, the methionines become hydrophilic and peptide **6** adopts preferentially a stable amphiphilic β-sheet conformation, thus allowing a redox-control over the conformation of **6**.

A similar peptidic conformational "switch" was developed by Mutter and Hersperger [9]. In that work, they showed that the 15-residue peptide **7** could have its conformation modulated by the polarity of the solvent (Fig. 6). Indeed, peptide **7** was shown to adopt an α-helical structure in trifluoroethanol and

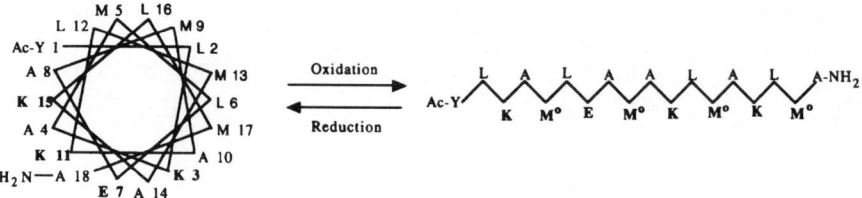

Fig. 5. Conformational modulation of a peptidic structure (**6**) controlled by a redox process [8]. The hydrophilic amino acids are in bold face and the sulfoxide form of methionine is denoted by M^O. On the left, the α-helix axial projection; on the right, the β-sheet side-view representation

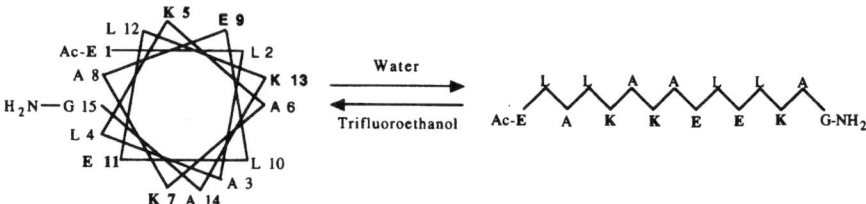

Fig. 6. Solvent-induced conformational changes in peptide **7** [9]. The hydrophilic residues are in bold face. On the left, the α-helix axial projection; on the right, the β-sheet side-view representation

a β-sheet conformation in water since in this polar solvent the β-sheet conformation forms a perfect amphiphilic structure with all the hydrophobic and hydrophilic residues on opposite sides of the backbone (Fig. 6).

Our group is also interested in developing peptidic structures that could have their conformation modulated by the specific recognition of effectors. In a first approach towards this goal, we designed and synthesized the short heptapeptides **8** to **10** that are composed only of alanines and two 18-crown-6 derivatives of phenylalanine [10]. Alanine was chosen because it can be accommodated in both α-helix and β-sheet structures, due to its small side-chain methyl group. On the other hand, the binding ability of crown ethers is well established. Our working hypothesis was that the backbone conformation could

Ac-E-L-A-L-K-A-K-A-E-L-E-L-K-A-G-NH$_2$ **7**

BOC-A$_3$-NH-C-CO-A-NH-C-CO-A-NHC$_3$H$_7$ **8**

BOC-A$_2$-NH-C-CO-A$_2$-NH-C-CO-A-NHC$_3$H$_7$ **9**

BOC-A-NH-C-CO-A$_3$-NH-C-CO-A-NHC$_3$H$_7$ **10**

R =

be modulated by the intramolecular binding of difunctional guests using the cooperative action of the two distant crown moieties, as shown schematically in Fig. 2. Depending on the distance between the crown ether residues and on the guest geometry, specific conformation should be induced and stabilized (Fig. 7). The side-chain binding cooperativeness of the bis-crown peptides **8–10** was first demonstrated by their high complexation ability towards difunctional guests such as Cs^+ and α, ω-alkyl diammoniums. Conformational studies using circular dichroism spectropolarimetry (hereafter CD) showed that the free peptides **8–10** exist in a β-sheet conformation in methanol, acetonitrile, and 1,2-dichloroethane. Upon addition of Na^+ and K^+ the conformation of the bis-crown peptides did not change as expected since these ions form one to one complexes with 18-crown-6 ligands. However, the addition of Cs^+ ions leads to an important conformational change in peptide **9** but not in peptides **8** and **10**. These ions are known to form sandwich complexes with one Cs^+ in between two 18-crown-6. It is conceivable that **8** and **10** have their binding side-chains well oriented for the cooperative complexation of Cs^+ in a β-sheet form. Therefore, no major conformational changes are required to orient the binding sites for the complexation process. However, contrarily to **8** and **10**, peptide

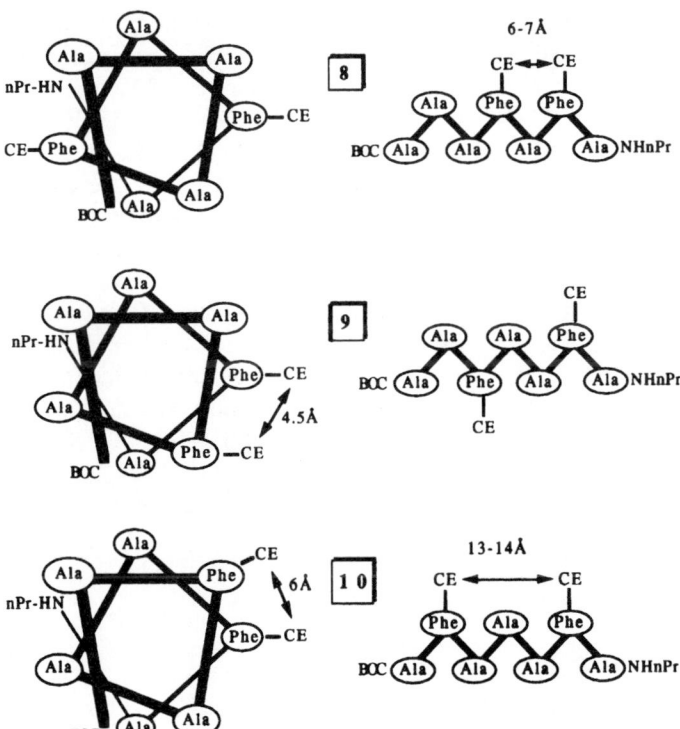

Fig. 7. Schematic representations of the α-helix (axial view) and the β-sheet (side view) conformations of peptides **8–10** (CE = crown-ether side chain). (Reproduced with the permission of Ref. 10a)

9 has the crown residues separated by two alanines and has the two crowns on opposite sides of the backbone in a β-sheet conformation. Therefore, peptide 9 needs an important conformational reorganization in order to orient the crown ethers in a complementary fashion to Cs^+. Interestingly, extensive CD studies suggest that the backbone of 9 switches from a β-sheet to a β-turn structure by the cooperative complexation of Cs^+, independently of the solvent used. This result is in contradiction with our original prediction that the binding of Cs^+ by 9 should induce and stabilize its α-helix conformation (Fig. 7). It is possible that 9 is too short to form a stable α-helix and that a β-turn structure also orients appropriately the two crown moieties for the complexation of Cs^+. Nevertheless, these results demonstrate the possibility of inducing stable β-turn structures using unnatural side chain interactions, independently of the nature of the solvent. In addition, CD studies with rigid diammoniums could help to demonstrate that the backbone conformation of these bis-crown peptides can be modulated by the recognition of guests with different geometries.

Using a related strategy, we investigated the possibility of developing peptide nanostructures with a defined conformation that could be modulated by a change in the pH [11]. Towards this goal, we synthesized peptides 11–13 that were designed to fold into a predictible conformation by the selective intramolecular complexation of a lysine γ-ammonium group by an 18-crown-6 modified lysine (Fig. 8). Here again, it was postulated that the unnatural interactions could induce a specific conformation, depending on the relative positions of the two residues that bear the binding partners (Fig. 9). CD studies in trifluoroethanol showed that 11–13 were mostly unordered and that the engineered interaction provided only a weak stabilization energy. The same studies in a less polar solvent (acetonitrile) demonstrated also that 11 and 13 were only slightly more conformationally stable than their analogs that lack the side-chain host-guest interaction, but mostly unordered. However, the same interaction made the β-sheet structure of peptide 12 notably more stable. Several control experiments demonstrated that the host-guest interactions were intermolecular between one ammonium group of a β-strand and a crown of another β-strand and stabilized the anti-parallel β-sheet structure of 12 as illustrated in Fig. 10. This result confirms that a heptapeptide is much more conformationally stable under a β-sheet form than under an α-helical one, the conformation that should

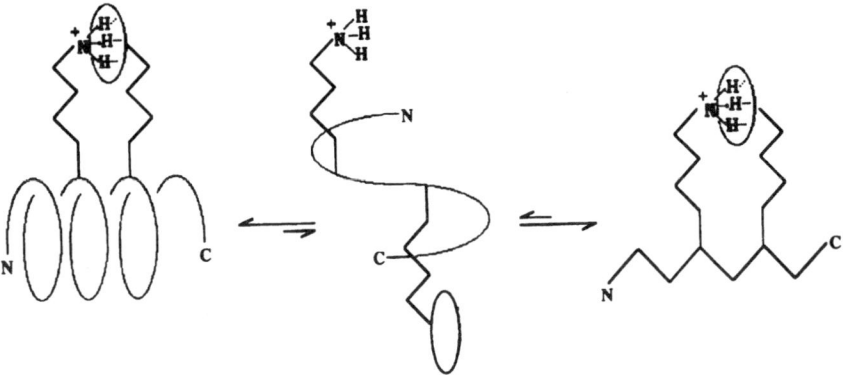

Fig. 8. Working hypothesis for the development of pH-gated peptide structures. Specific conformations can be induced and stabilized by the formation of a side-chain host-guest complex between an ammonium group and a crown ether. (Reproduced with the permission of Ref. 11).

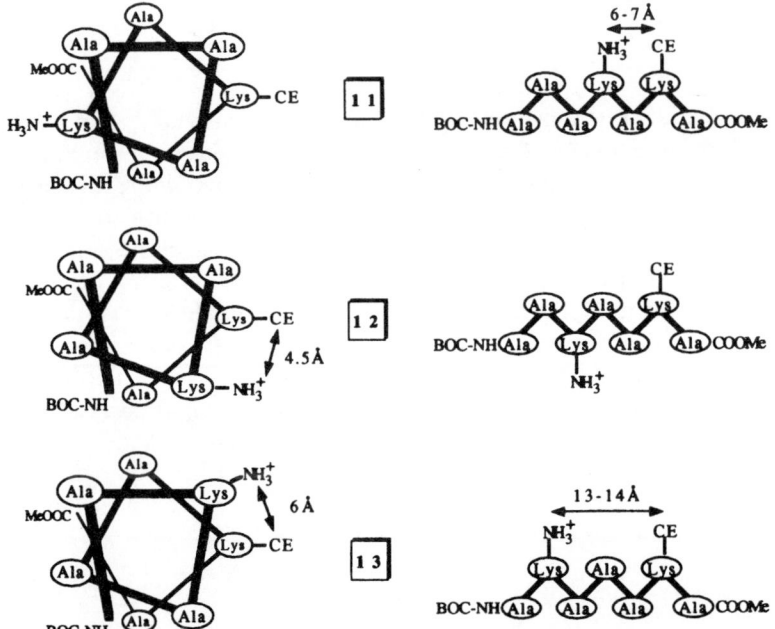

Fig. 9. Schematic representations of the α-helix (axial view) and the β-sheet (side view) conformations of peptides **11–13** (CE = crown-ether side chain). (Reproduced with the permission of Ref. 11).

have been stabilized by an intramolecular host-guest interaction in **12** (Fig. 9). It also illustrates clearly that subtle changes in peptidic structures, such as the relative positions of two interacting residues, can have a profound influence on their conformational stability in solution and that more work is necessary to be able to predict these influences adequately.

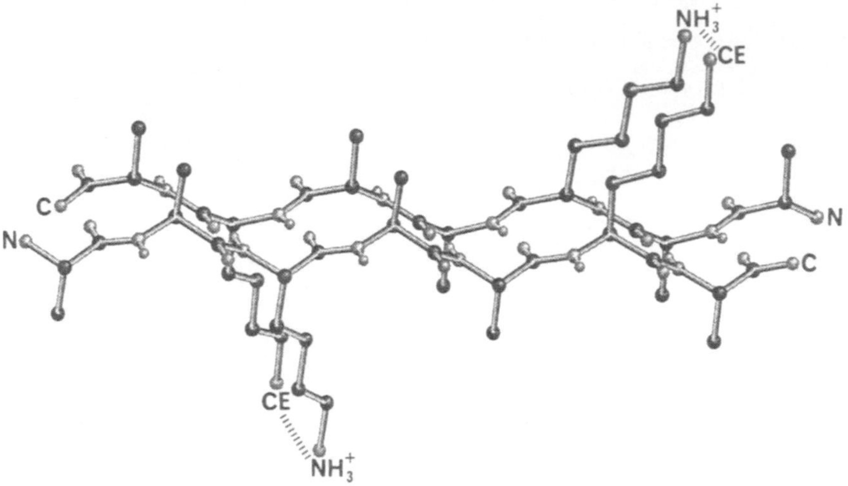

Fig. 10. Portion of the β-sheet structure of peptide **12** in acetonitrile. The structure is stabilized by interstrand host-guest interactions. (Reproduced with the permission of Ref. 11)

The conformational stabilization of β-turn structures is also an area of great importance. Not only because β-turn structures are connectors that link and orient correctly the other secondary structures to create the globular, active form of proteins and enzymes, but also because well defined β-turn structures are involved in numerous important biological recognition events. In this respect, Fisher and Imperiali [12] reported the preparation of a well defined β-turn peptide structure **14** in which the conformational stability is ensured by the formation of a Cu^{2+} complex between two distant bipyridine modified alanines. The complex was structurally defined and the Cu^{2+} was found to be complexed in a chiral environment. The same group also reported [13] the stabilization of β-turn structures of heptapeptides **15** and **16** by the formation of well defined Zn^{2+} complexes with either the carboxylate of the glutamic acid and the imidazole of the histidines at positions 6 and 8 in **15** or the three imidazoles of the histidines of **16**. Interestingly, with **17**, the analogue of **16** that

Ac-**Bpa**-T-P-(D)A-V-**Bpa**-NH$_2$ **14**   Bpa= -NH-C-CO-

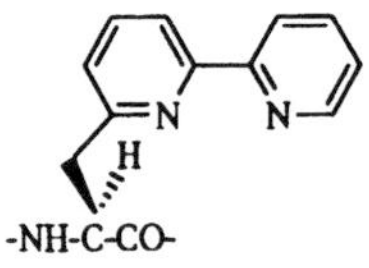

Ac-**E**-G-V-P-(D)S-**H**-T-**H**-NH$_2$ **15**

Ac-**H**-G-V-P-(D)S-**H**-T-**H**-NH$_2$ **16**

Ac-**H**-G-V-G-G-**H**-T-**H**-NH$_2$ **17**

lacks a β-turn nucleation site (Pro-D-Ser replaced by a more flexible Gly-Gly sequence), it was found that the metal complexation process alone was not sufficient to force the backbone to adopt a well defined β-turn structure.

Another way of stabilizing a specific secondary structure is to form somewhat permanent covalent bonds either between two different side chains or between the N- and C-terminal positions. The latter have been used extensively [14] in the development of rigid β-turn structures and will not be reviewed here. The formation of stable α-helical peptides by the formation of side-chain covalent bonds has been reported by Ösapay and Taylor [15]. They synthesized a peptide 18 designed to be "locked" into an helical conformation by the formation of several amide bonds between glutamic acid and lysine side chains located three residues apart. Peptide 18 was shown to be highly helical and the secondary structure was very stable to denaturation. In addition, the same group recently reported the preparation of the highly α-helical hexapeptide 19 by the same strategy [16]. Indeed, peptide 19 having two side-chain lactam bridges between lysines and aspartic acids, separated by three residues, was shown by CD and NMR to exist in an α-helical conformation in a trifluoroethanol/water (1/1) solution. It is noteworthy that the peptide partially retains this conformation even at 70 °C.

Along the same lines, Schultz and co-workers [17] have used the formation of an intramolecular disulfide bridge to induce and enhance the α-helical structure of alanine-based peptides such as 20. They designed and synthesized peptides so that in a helical conformation two side chain thiol groups are oriented correctly to form an intramolecular disulfide link that is meant to lock the backbone in that conformation. Using this strategy, the helical content of the octapeptide 20 increased from 16% to roughly 100% at 0 °C in water when forming the disulfide bridge. The helical content remained high (59%) even at 60 °C. They also showed with a longer analog 21 that the helical conformation could be propagated to adjacent amino acids resulting in a totally helical conformation in 21 in water at 0 °C. They also noted that the use of the L-enantiomer of the thiol modified amino acid at position 1 of 20 instead of the D led only to a modest increased in the helicity (31%). This result illustrates once again that subtle changes can have an important influence on the conformational stability of peptidic nanostructures. The formation of covalent disulfide bonds has also been used to stabilize the three dimensional structure of artificial α-helix bundle proteins. For instance, Chmielewski and Lutgring [18] formed

$$\text{HN}\overline{}\text{CO}$$
H-(K-L-K-E-L-K-D)$_3$-OH **18**   BOC-K-K-A-A-D-D-OPac **19**
$$\text{HN}\overline{}\text{CO}$$ $$\text{HN}\overline{}\text{CO}$$

$$\text{S}\overline{}\text{S}$$
Ac-X-K-A$_4$-K-X-NH$_2$ **20**   H ,(CH$_2$)$_4$-

Ac-A$_3$-X-K-A$_4$-K-X-A$_3$-K-A-NH$_2$ **21**   X= -HN-C-CO-
$$\text{S}\overline{}\text{S}$$

11

disulfide bonds between helices of a self-assembled protein and observed an increase in its structural stability. Contrary to their expectations, using the amphiphilic peptide unit **22** they obtained a covalently stabilized structure containing five helices rather than the usually more stable structure with four helices.

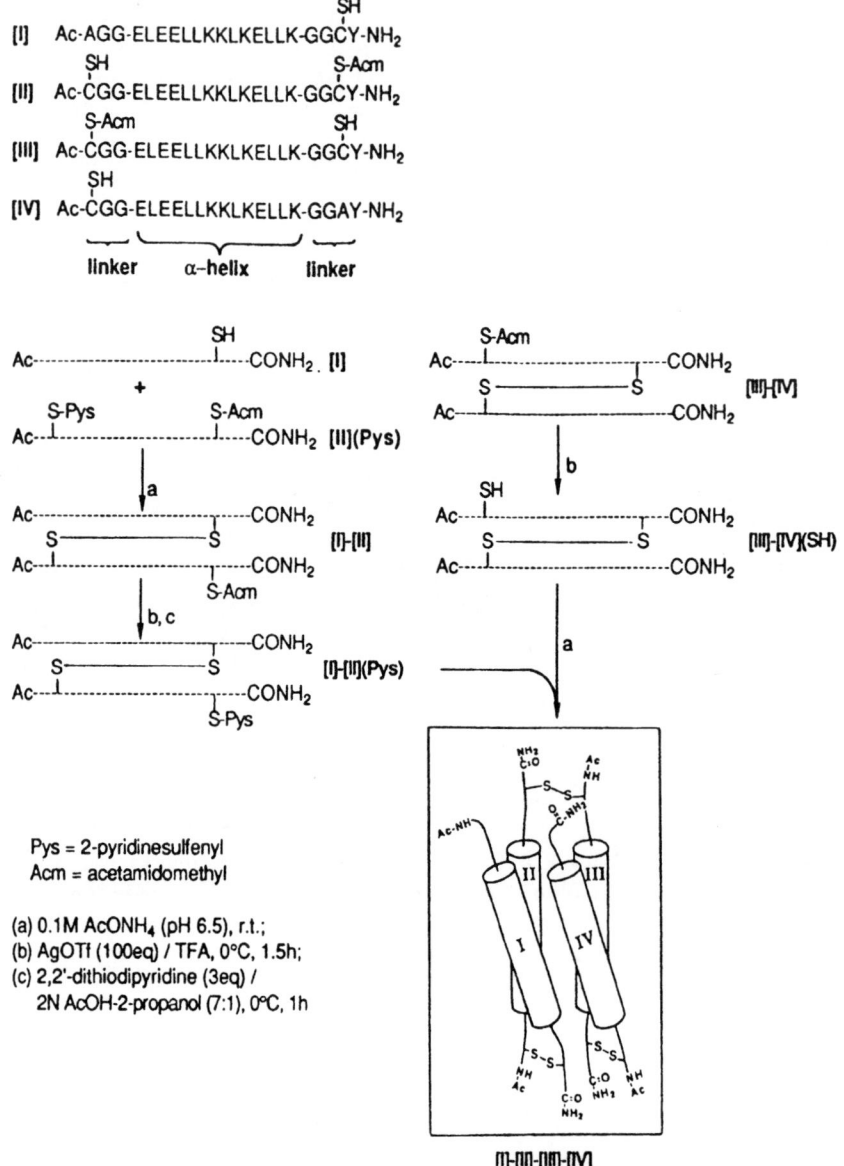

Fig. 11. The preparation of a four α-helix structure by cross linking four different peptide units (I, II, III, IV) through disulfide bonds. (Reproduced with the permission of Ref. 19)

$$CH_2SH$$
$$H$$
Ac-E-L-L-K-L-X-K-X-L-E-E-L-K-K-NH$_2$ **22**   X= -HN-C-CO-

On the other hand, Futaki and Kitagawa [19] prepared a four-helix bundle protein that was also stabilized by disulfide bonds. However, in this case, they synthesized the artificial protein with its four different peptide structures (I, II, III, IV) by selectively forming the disulfide bonds, as illustrated in Fig. 11.

Finally, extending the concept of helix capping, developed by Rose [20] as well as by Richardson [21], several groups have demonstrated that the helicity of short peptides can be significantly increased by adequately capping the C- and N-terminal positions with synthetic groups that can form hydrogen bonds with the NH or CO terminal groups and interact favorably with the helix dipole. For example, Nambiar and coworkers [22] showed recently that the helicity of a 16-residue peptide (**23**) could be increased from 17% for the free NH$_2$ peptide to 73% in water, simply by acylating the N-terminal position with an alkylsulfonic acid group (**24**). This strategy could be used, in combination with the others presented above, to develop quite stable and structurally defined peptides.

X-A$_4$-E-A$_3$-K-A$_4$-Y-R-NH$_2$ **23**   X=H   **24**   X= HOSO$_2$CH$_2$CH$_2$CO-

2.2 Using Templates and Nucleating Devices

The use of templates that can nucleate secondary structures has also been studied [23]. The fundamental idea is to attach one or more conformationally flexible peptides to a rigid template that is designed to initiate either a β-sheet or an α-helix by forming the first crucial hydrogen bonds. These interactions compensate for the loss of entropy associated with the folding process and in particular in the initiation step. This strategy has been used to develop stable helices, sheets, and artificial proteins.

An important contribution to the nucleation of α-helices has been made by Kemp and co-workers [24]. They prepared a rigid template (**25**) that was designed to induce stable helices by forming three complementary hydrogen bonds with a peptide attached by its N-terminal group, as illustrated in Fig. 12. Extensive NMR and modelling studies demonstrated that the helical template **25a**, the free acid, is in conformational equilibrium between three states, one of them being a nucleation state. It was also shown that **25** effectively nucleates and stabilizes the α-helical conformation of alanine based peptides. Interestingly, they used **25** to re-evaluate the s-constant (the intrinsic propensity to induce a helical structure) of alanine in water [24a]. They found an s-constant of 1 for alanine which agrees with the one calculated by Scheraga and co-workers.

The nucleation of β-sheets has been studied by the groups of Kemp, Feigel, Kelly, and Nowick. The former designed and synthesized the C$_2$-symmetric

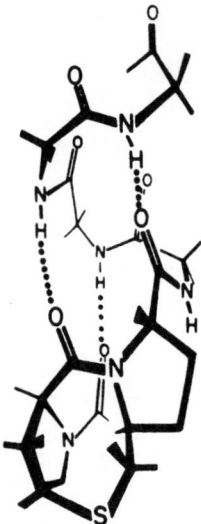

Fig. 12. Templating α-helices using compound **25** [24b]. The template (bottom part) serves to initiate the helical folding by forming the first hydrogen bonds. (Reproduced with the permission of Ref. 24b)

diamino compound **26** with the aim of facilitating the formation of β-sheets in two peptides attached to the amino groups by forming the complementary hydrogen bonds (Fig. 13) [25]. Using NMR, they showed that **26** adopted the proposed unaggregated β-sheet structure depicted in Fig. 13 in DMSO. Towards the same objective, Feigel and co-workers [26] synthesized a series of compounds, such as **27** and **28**, with the aim of substituting a β-turn and orienting two peptide chains to form parallel and anti parallel β-sheets, as shown, for example, in **29**. Even though they prepared only cyclic peptides, they

25 a) X= OH   b) X= -A$_n$-OH   n= 2-7

26

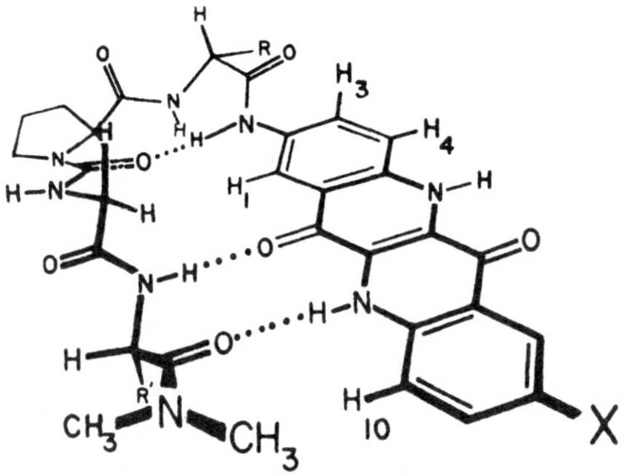

Fig. 13. Illustration of the β-sheet structure generated by the template compound **26**. (Reproduced with the permission of Ref. 25)

demonstrated by NMR and molecular modelling that the peptidic chains adopted a β-sheet conformation.

Inspired by cyclic peptides such as Gramicidin S, Kelly and co-workers [27] designed a dibenzofuran template (**30**) that orients the two peptides in the proper arrangement to form a stable two-stranded anti-parallel β-sheet structure. They showed that **31** adopted a partial β-sheet conformation, whereas its acyclic analog (**33**) having the template substituted by the D-Phe-Pro unit of Gramicidin S was structureless in water. They extended their work to prepare **32**, which adopted a strong homogeneous β-sheet conformation in water. More

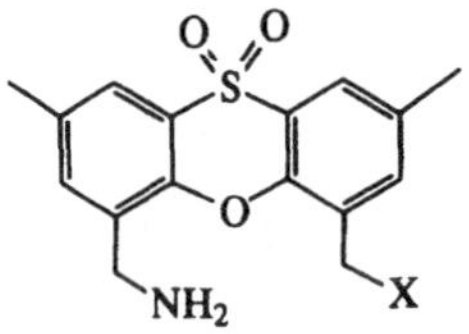

27 X= NH₂ or COOH

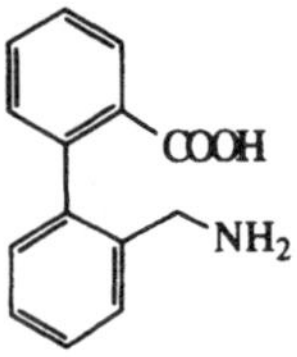

28

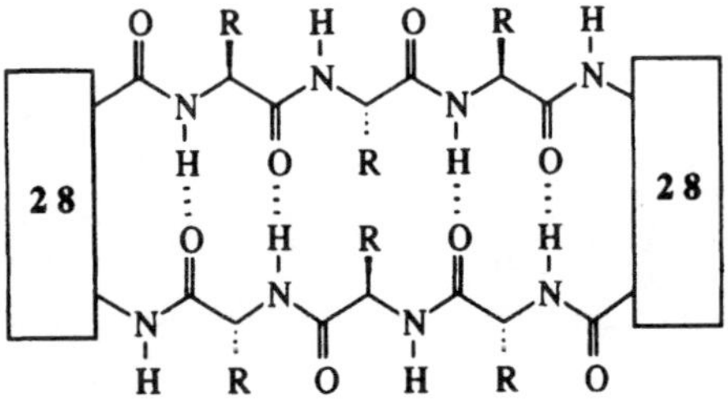

29

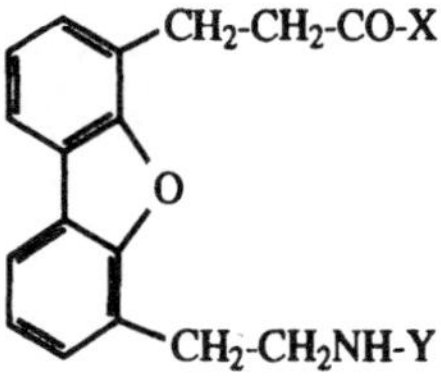

30 X= OH   Y= H

31 X= -V-K-L-NH₂   Y= -L-K-V-NH₃⁺

32 X= -V-K-V-K-V-K-NH₂   Y= -V-K-V-K-V-K-NH₃⁺

X = Duplicate of peptide portion

15

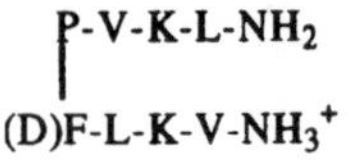

P-V-K-L-NH₂

(D)F-L-K-V-NH₃⁺

33

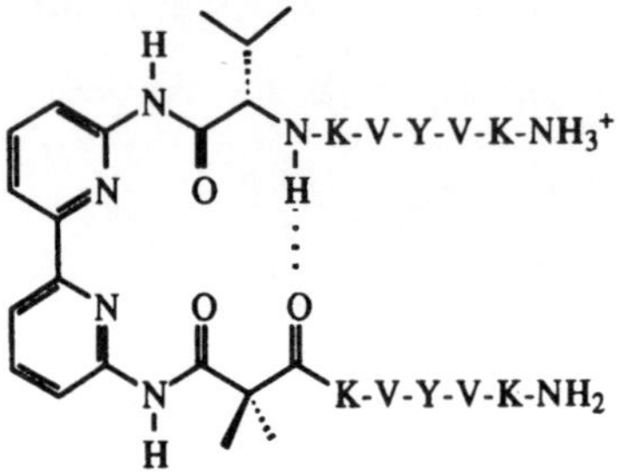

34

recently, they developed an elegant system **34** that adopts a well-defined β-sheet structure upon complexation with Cu²⁺ [28]. Indeed, the binding of copper to **34** switches the bipyridyl unit from the *trans*- to the *cis*-conformation, which orients the two peptidic units in close proximity where they adopt a β-sheet structure (Fig. 14).

On the other hand, the group of Nowick [29] developed polyurea templates with the general structure **35** in order to nucleate parallel β-sheet structures. They reported that in a model compound (**36**) the urethane carbonyls are hydrogen bonding to the adjacent NH and that the orientation of the bonding interactions is controlled by the size of the end substituent, in this case a phenyl group. They also recently prepared **37** and showed by NMR that it adopts mostly the proposed parallel β-sheet conformation in chloroform [30].

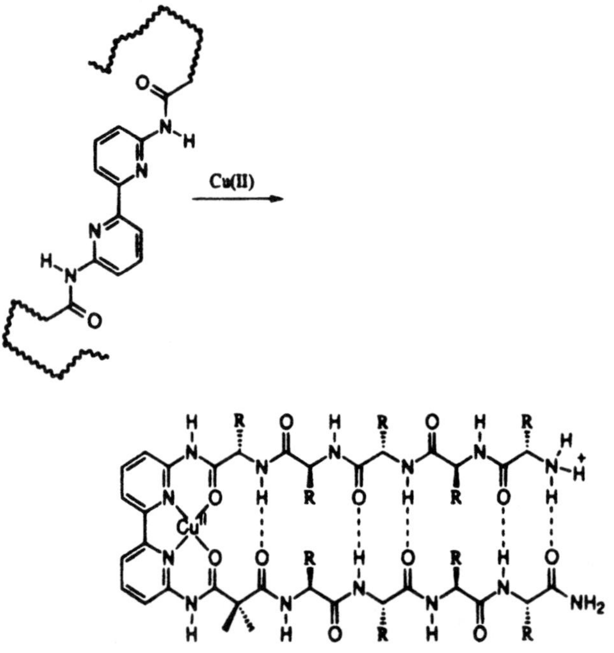

Fig. 14. The formation of a double-strand β-sheet structure, induced by the binding of Cu(II). (Reproduced with the permission of Ref. 28)

35

36 $R_1 = R_2 = Ph$

37 $R_1 = $ -F-L-NH-Me
$R_2 = $ -V-A-NH-Me

The preparation of structurally defined protein-like structures has also been realized by attaching several amphiphilic peptide segments to rigid templates that bear multiple functional groups. Pioneering work in this area was done by Mutter who developed the TASP concept (Template-Assembled Synthetic Proteins) [31]. Here, the molecular scaffold holds the peptidic chains close to each other and favors the first interchain contacts, thereby overcoming the loss of entropy due to the folding process. This strategy therefore allows *de novo*-designed artificial proteins to be synthesized that have a predetermined structure and, conceptually, all possible topologies (Fig. 15). Mutter and his co-workers [32] had already reported the preparation and characterization of templated proteins with a four α-helix bundle structure, such as **38**, using

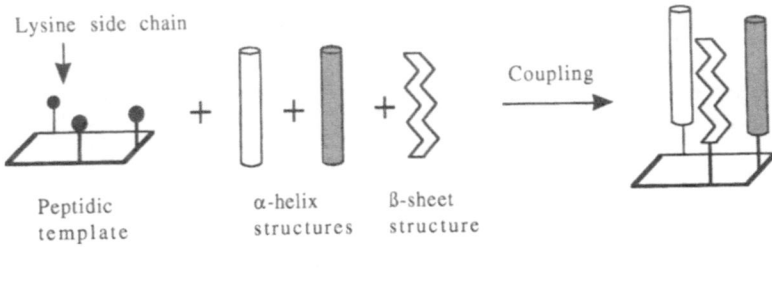

H₂ NOC-Cys-Lys-Ala-Lys-Gly
      S–S
AcNH-Cys-Lys-Ala-Lys-Pro

38 ● = -G-L-K-A-L-A-E-K-L-A-K-E-L-A-E-NH$_3^+$

Fig. 15. The construction of artificial proteins by the TASP (template-assembled synthetic proteins) concept [31]. Peptidic structures are attached to a synthetic template to create artificial proteins with a defined three-dimensional architecture. (Reproduced with the permission of Ref. 1)

peptidic templates. Even though the TASP strategy has not led to the preparation of artificial proteins incorporating β-sheets and catalytic functions, this approach to the development of peptide nanostructures clearly has great potential. This is confirmed by the synthesis of a TASP molecule with ion-channel activity by Montal and co-workers [33].

Other templates have been described in the literature. Three independent groups used a porphyrin template. Sasaki and Kaiser utilized a porphyrin bearing four carboxylic acids (39) to prepare a synthetic metalloprotein with a four α-helix topology (40) [34]. The utilization of a porphyrin template allows not only the templation of amphiphilic peptide segments, but also the incorporation of a metal binding site, which could serve eventually in a catalytic process (Fig. 16).

A porphyrin template has also been used by DeGrado and coworkers [35] for developing a well-defined four α-helix structure (41) that has proton channel activity (see Sect. 3.2). Following modeling studies, they used a tetraphenyl porphyrin with four carboxylic acids at the meta positions as attachment points for the peptide segments.

The ortho- and para carboxylic acid substituted tetraphenyl porphyrins were used by Nishino and coworkers [36] to prepare the well defined 4-α-helix structures 42 and 43, which are soluble in lipid bilayer membranes. The rigid template of 42 provides some stability to the bundle structure (as measured by

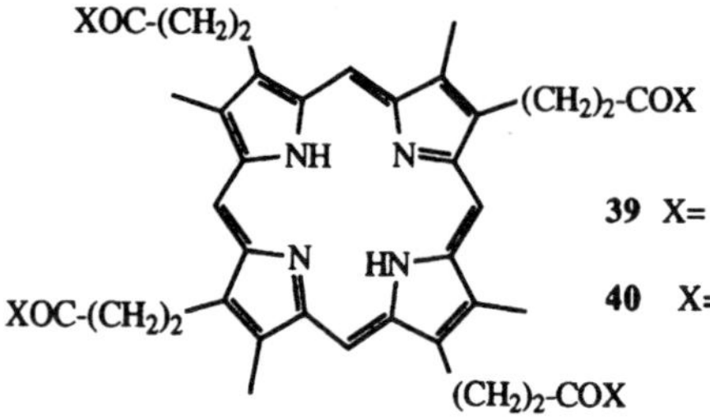

39 X= OH

40 X= -A-E-Q-L-L-Q-E-A-E-Q-L-L-Q-E-L-NH₂

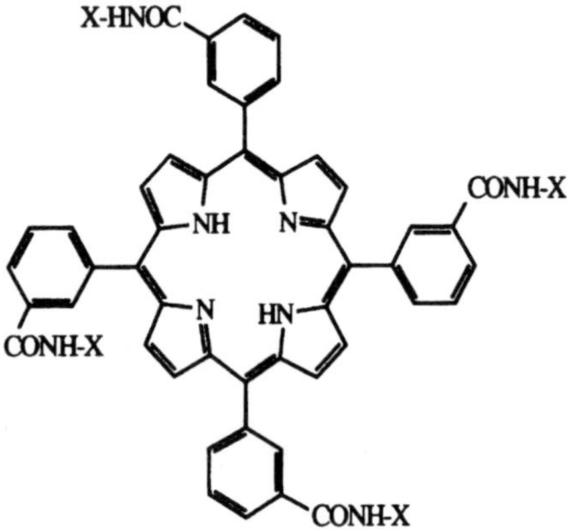

41 X= -(L-S-Aib-L-S-L)₃-NH₂

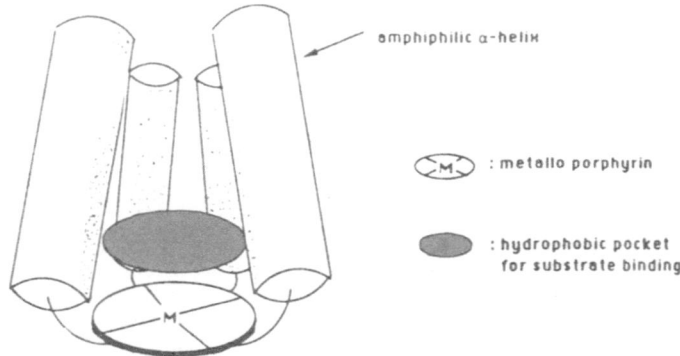

Fig. 16. Schematic representation of a four α-helix structure (**40**) showing the metalloporphyrin and the hydrophobic pocket where a small aromatic substrate can be complexed [34]. (Reproduced with the permission of Ref. 34b)

42 X= -CO-A-(A-L-L-E-L-L-A)₃-NHMe
Z= -H

43 X= -H
Z= -CO-A-(A-L-L-E-L-L-A)₃-NHMe

CD) in comparison with the star shaped analog **43**. They also showed that **42** is incorporated easily into phospholipid bilayers. On the other hand, the same group reported the preparation of the coiled coil structure **44** with two amphiphilic 14-residue peptides linked to a bipyridyl template [37]. The incorporation of a fluorescent probe, pyrenyl-L-alanine, near the supercoiling region helped them to demonstrate by fluorescence the formation of the proposed structure in water.

Recently, a modified β-cyclodextrin was also used as a scaffold in the preparation of a structure that has seven β-sheet-forming peptides (**45**) [38]. Although no structural analysis was reported, this represents the first synthesis of a peptidic architecture with seven adjacent peptide units.

Finally, Sherman and co-workers [39] recently reported the preparation of a four α-helix structure **46** by attaching the peptide units onto a synthetic bowl

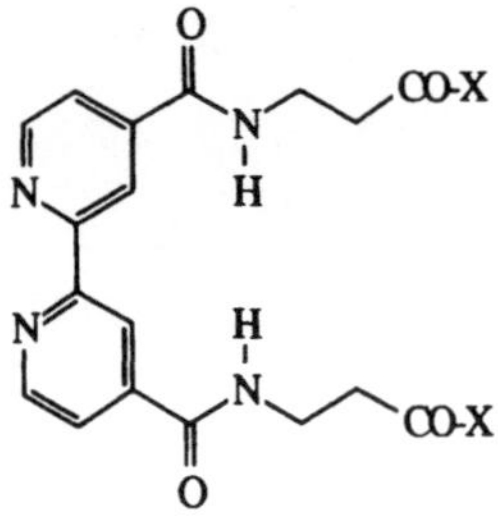

44 X= -A-L-A-K-**Pya**-A-A-L-A-K-A-L-A-NH$_2$

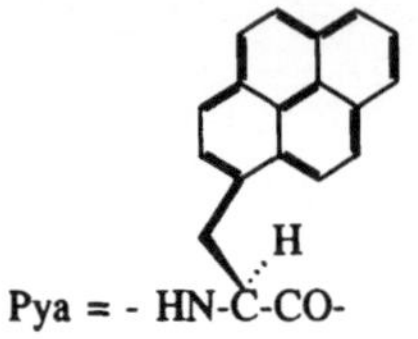

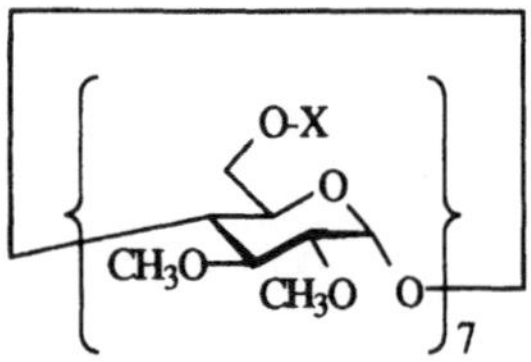

Pya = - HN-C-CO-

45 X= -CH$_2$-CONH-W-S-L-S-L-S-L-NH$_2$

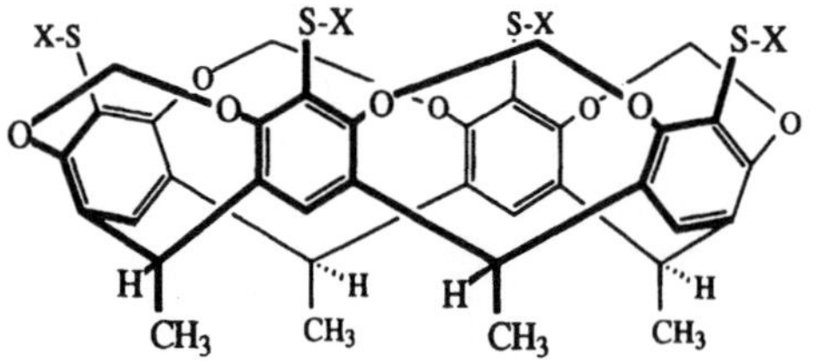

46 X= -CH$_2$CO-E-E-L-L-K-K-L-E-E-L-L-K-K-G-NH$_2$

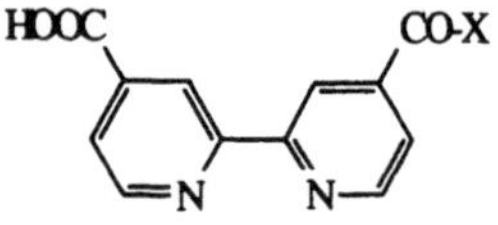

47 X= A-E-Q-L-L-Q-E-A-E-Q-L-L-Q-E-L-NH$_2$

template. Just as in the case of the cyclodextrin template, one advantage of this approach is the incorporation of a binding pocket (template) in the artificial protein. The major disadvantage, however, is the number of topologies that can be prepared by this strategy is limited by the synthesis and the geometry of the templates.

Using a different approach, Lieberman and Sasaki [40] described the formation of a well-defined triple helix structure by the self-assembly of three peptide segments (**47**) via the formation of an iron (II) bipyridyl complex (Fig. 17). Independently, Ghadiri and Case [41] also reported the preparation of a triple helix structure using a similar strategy with peptide **48**. In this case, the template is formed by the coordination of Ni(II), Co(II), or Ru(II) with three bipyridine units. The triple helix structure with Ru(II) was characterized by mass spectrometry. Furthermore, CD studies showed an increase in helicity and structural stability. By a clever modification, the same authors used the self-

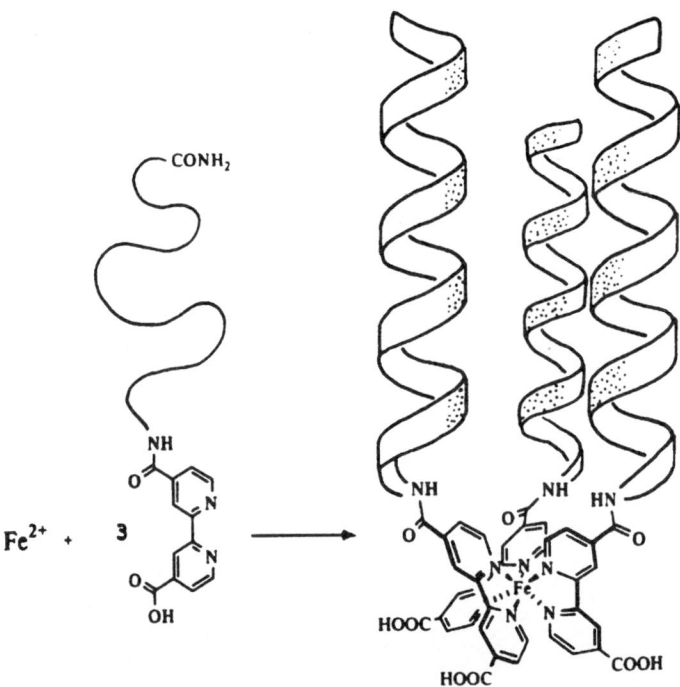

Fig. 17. Illustration of the self-assembly of an artificial triple-helix structure mediated by the complexation of Fe(II) with three bipyridine modified peptide **47**. (Reproduced with the permission of Ref. 40a)

—CONH-G-E-L-A-Q-K-L-E-Q-A-L-Q-K-L-A-X

48 X= -NH₂   **49** X= -A-A-H-Y-NH₂

assembling process to create a binding site for another metal by accurately orienting the imidazole groups of three histidines near the C-terminal of **49**, as depicted in Fig. 18 [42].

They showed that Cu(II) was bound in that position and that the binding process did not influence the tertiary structure. Moreover, the binding of Cu(II) further enhances the structural stability by 1.5 Kcal/mol. In a similar way, the group of Ghadiri [43] has succeeded in preparing a well defined four α-helix structure **50** by forming a Ru(II) complex with four pyridine-modified peptide units **51**.

With somewhat the same idea, Minoura [44] used the formation of a crown ether sandwich complex with K⁺ to force the association of helical dimers of C-terminal modified poly-benzylglutamate (**52**). Although no structural data were provided, he observed by viscosimetry the reversible formation of a K⁺ sandwich complex with two 15-crown-5 modified peptides.

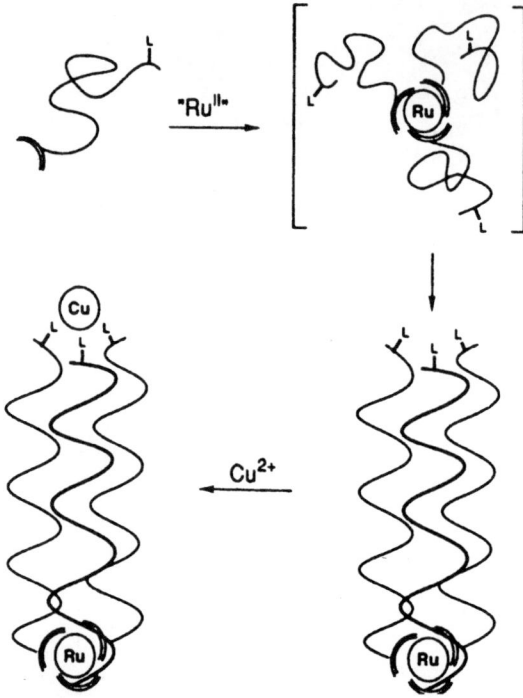

Fig. 18. Induced by the binding of Ru(II), the self-assembly of three peptide units **49** creates another binding site for Cu(II). (Reproduced with the permission of Ref. 42)

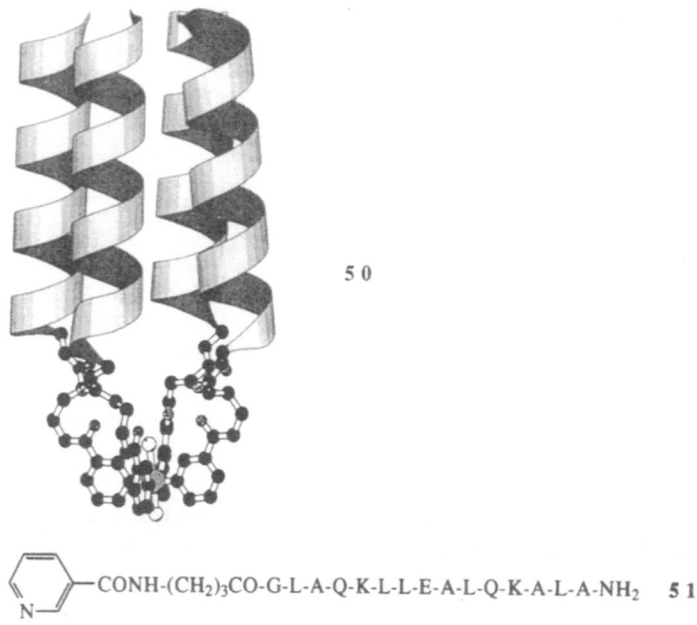

5 0

—CONH-(CH₂)₃CO-G-L-A-Q-K-L-L-E-A-L-Q-K-A-L-A-NH₂   5 1

22

The above results illustrate the diversity of unnatural tricks that chemists have used recently in order to help peptidic molecules adopt a well defined structure in solution. Efforts were directed to the development of both single and multiple peptide structures.

3 Functional Peptide Nanostructures

The ultimate goal is to be able not only to develop structurally defined peptide nanostructures, but also to render them functional by incorporating a binding, a reactive, or a catalytic site. The term functional here is used in a very broad sense. In the following lines, we shall illustrate the potential of peptide nanostructures in miscellaneous areas.

3.1 Molecular Receptors

Our research group and others have recently been interested in developing molecular receptors for biologically important compounds by taking advantage of peptides as frameworks. For instance, we and others have used crown ether modified peptides to develop molecular receptors that can complex (and potentially extract) selectively certain ions such as Cs^+. As mentioned above, peptides **8–10** demonstrated a strong binding selectivity towards Cs^+ [10]. On the other hand, Berthet and Sonveaux [45] showed that receptors such as **53** with enough flexibility were able to complex selectively with K^+. Interestingly, the same authors [46] prepared the poly crown L-phenylalanines **54** and **55** and investigated their chiral recognition ability towards leucine methyl ester. However,

53 X= Ala; Pro; Pro-Ala; Ala$_3$; Ala-Pro-Ala; -NH-(CH$_2$)$_7$- CO-

54 X=1 **55** X=2

H-(-NH-C̈-CO-)$_n$-OH with COOR, H

56 R= a mixture of -H, -CH$_2$-Ph, and

57 R= a mixture of -CH3 (66-86%) and

-CH$_2$-(CH$_2$-O-CH$_2$)$_x$-CH$_2$-O—⟨aromatic⟩—(CH$_2$)$_8$-CH$_3$

X = 7.5, 10, 15 (14-34%)

BOC-A-X-A$_2$-X-A-OMe   **58**

BOC-A-X-A$_3$-X-A-OMe   **59**

BOC-A-X-A$_6$-X-A-OMe   **60**

X= -NH-C̈-CO-

these polypeptides did not discriminate between the two enantiomers. The same results were observed by Osa and Ueno [47] using a polyglutamate in which some side chains had been modified with benzo crown ethers (**56**) or with polyethylene glycol (**57**).

Polyamines are biological molecules of interest and the development of molecular receptors for these compounds has attracted considerable attention. In order to complex selectively certain diamines, we have designed and synthesized the bis porphyrin peptides **58–60** [48]. Of these receptors, only the shorter one (**58**), which has the two porphyrin amino acids separated by two alanines, exhibited a strong binding selectivity for 1,5-diaminopentane in a binding study with diaminoalkanes (Fig. 19). This selectivity is due to the cooperative action of the two zinc porphyrins in the complexation of diaminopentane. It is noteworthy that the longer and more flexible receptors **59** and **60** do not complex the diaminoalkanes by using the two binding side chains. It is conceivable that these peptidic receptors could selectively bind more rigid substrates that have two more distant amino groups.

Peptidic receptors, targeted for the complexation and transport of biogenic amines, have also been developed in our laboratory [49]. The design of receptors **61–63** was inspired by the binding site models of dopamine and related amines natural receptors [50]. The crown ether was chosen to mimic the hydrogen bonding and electrostatic binding site and a flexible benzyloxycarbonyl lysine was chosen to mimic the aromatic interactions site, as illustrated in Fig. 20. Receptors **61** and **63** (but not **62**) were shown to complex more tightly aromatic ammonium in chloroform, using the proposed interactions. Extensive conformational studies by CD and NMR spectroscopy confirmed that the receptor framework existed in a β-sheet conformation in which only **61**

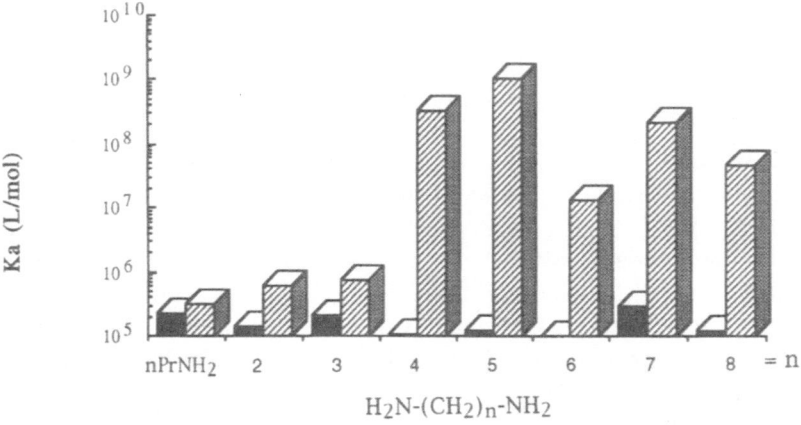

Fig. 19. Binding constants observed for the complexation of diamines by the bis-zincporphyrin peptide **58** (striped columns) and a mono-zincporphyrin tripeptide analog (BOC-A-Zn·porphyrin-E-A-OMe) (solid columns). (Reproduced with the permission of Ref. 1)

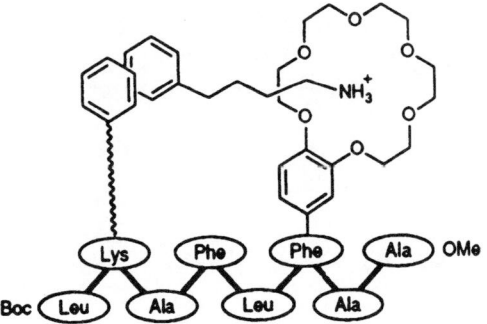

Fig. 20. Proposed binding mode for the complexation of an aromatic ammonium guest by peptidic receptor **63**. (Reproduced with the permission of Ref. 49)

25

and **63** have their two binding side chains oriented for complexation. In addition to the development of selective transporters of biogenic amines, these studies demonstrated experimentally the importance of aromatic-aromatic interactions in the complexation of these neurotransmitters.

In biological recognition phenomena, protein-protein interactions are of primary importance. In an attempt to mimic these processes, LaBrenz and Kelly [51] synthesized the peptidic host **64**. In this receptor, the dibenzofuran template separates the two peptide units by roughly 10 Å and allows for the complexation of a guest peptide (**65**), as depicted in Fig. 21. The complex first forms a three-stranded, antiparallel β-sheet that is stabilized by hydrogen bonds, electrostatic interactions, and aromatic-aromatic interactions between the dibenzofuran and the benzamide moieties. This complex can further self associate to form more complex structures. This example shows that structurally defined peptide nanostructures can interfere with biological recognition processes and potentially have therapeutic applications.

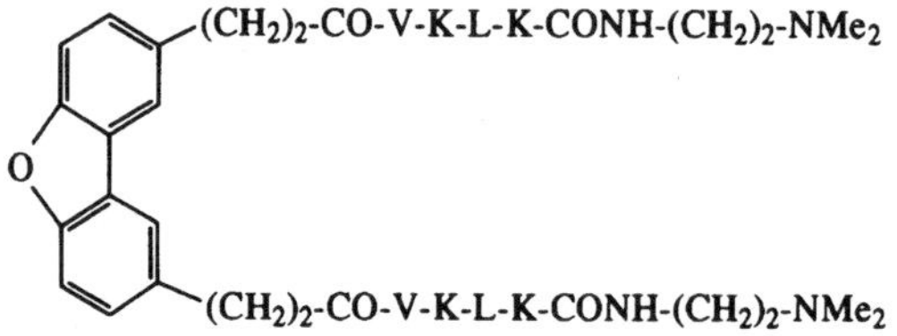

6 4

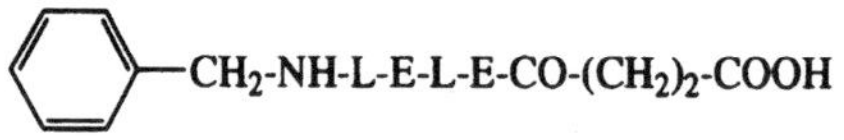

3.2 Ion Channels

Ion channel proteins are large and complex membrane proteins that regulate the flow of ions across cell membranes [52]. Since they are associated with several diseases, major efforts have been devoted to the preparation of functional artificial ion channels [53]. The challenge here is to design simple

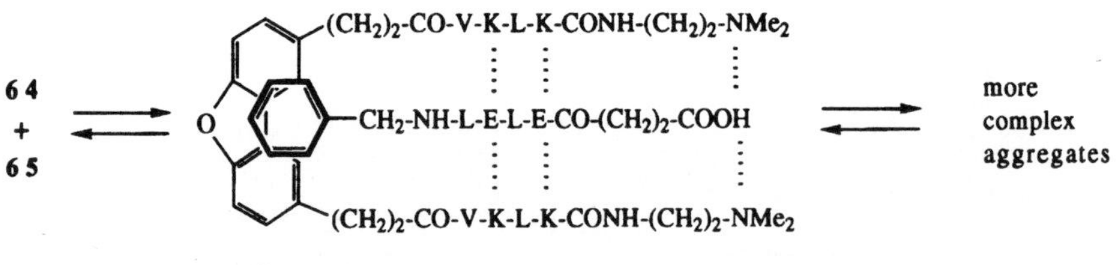

Fig. 21. Binding equilibrium between host peptide **64** and guest peptide **65** [51]

H-(L-S-S-L-L-S-L)$_3$-NH$_2$ **66**

$$\underset{\text{Ac-}\overset{|}{\text{K}}\text{-}\overset{|}{\text{K}}\text{-}\overset{|}{\text{K}}\text{-P-G-}\overset{|}{\text{K}}\text{-E-}\overset{|}{\text{K}}\text{-G-OH}}{\overset{\text{X X   X X}}{}}$$ **67**

X= $^+$H$_3$N-E-K-M-S-T-A-I-S-V-L-L-A-Q-A-V-F-L-L-L-T-S-Q-R-CO-

functional channels so as to be able to understand their mechanisms. A seminal contribution has been made in this field by DeGrado and co-workers [54]. Inspired by the natural ion channel proteins, their approach was to design amphiphilic peptides that could self-assemble into hydrophobic four α-helix bundle structures to form artificial ion channels, as illustrated in Fig. 22. Although they succeeded in preparing functional channels by aggregating peptide structures such as **66**, the synthesis, the purification, and the characterization turned out to be difficult especially in determining the active membrane structure. Using this work as their basis, DeGrado and Groves [35] showed that the four α-helix structure **41** acts as a more active H$^+$ channel than the peptide unit itself. They attributed this result to the porphyrin template that stabilizes the active 4-helix bundle structure of the channel.

Along the same lines, an artificial ion channel was prepared by Montal and coworkers [33] using the TASP approach. In their work, a four α-helix bundle structure **67** was synthesized on a peptide template. The ion transport ability was well characterized and **67** turned out to have several similarities with the natural acetylcholine receptor channel they were mimicking.

In a different approach to the development of an artificial channel, Ghadiri and coworkers [55] designed a cyclic octapeptide (**68**) composed of amino acids with alternating D and L configurations. The resulting flat macrocycle was

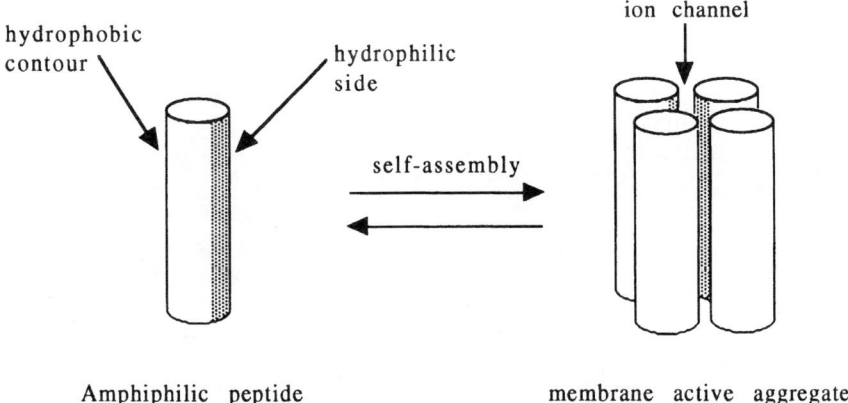

Fig. 22. DeGrado's concept for the development of artificial ion channels. The self-assemblage in a bilayer membrane of an amphiphilic peptide unit generates a hydrophobic α-helix bundle structure with a polar channel in the middle

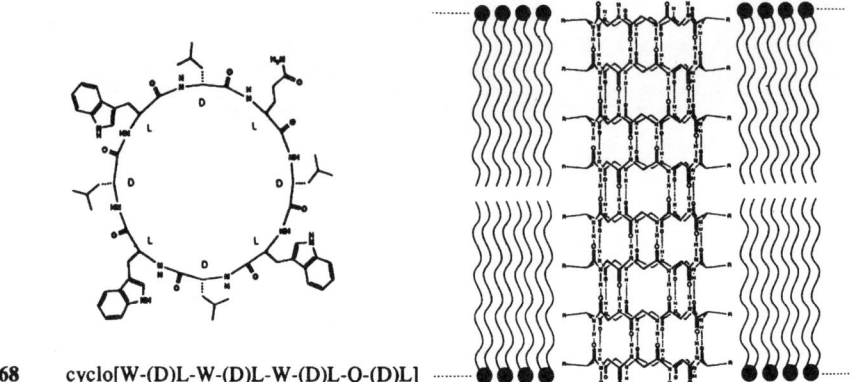

68   cyclo[W-(D)L-W-(D)L-W-(D)L-Q-(D)L]

Fig. 23. Schematic representation of the transmembrane channel created by the self-assembly of eight cyclic peptides **68**. (Reproduced with the permission of Ref. 55)

BOC-(L-HN-$\overset{\overset{\displaystyle X}{|}}{\underset{\underset{\displaystyle H}{}}{C}}$-CO-L$_3$-HN-$\overset{\overset{\displaystyle X}{|}}{\underset{\underset{\displaystyle H}{}}{C}}$-CO-L)$_3$-OMe   **69**   X=

shown to form nanotubes by self aggregation. Addition of **68** in a planar lipid bilayer resulted in membrane conductivity similar to the one observed with ion channels and the authors proposed that the channel is formed by the assemblage of eight macrocycles as illustrated in Fig. 23. They also extended their work to the development of a glucose transport system by the same strategy using a larger cyclic peptide [56].

Finally, our approach to the development of artificial ion channels is based on the alignment of six crown ethers attached to an α-helical peptidic nanostructure (**69**) (Fig. 24) [53, 57]. Preliminary results demonstrated that **69** adopts the

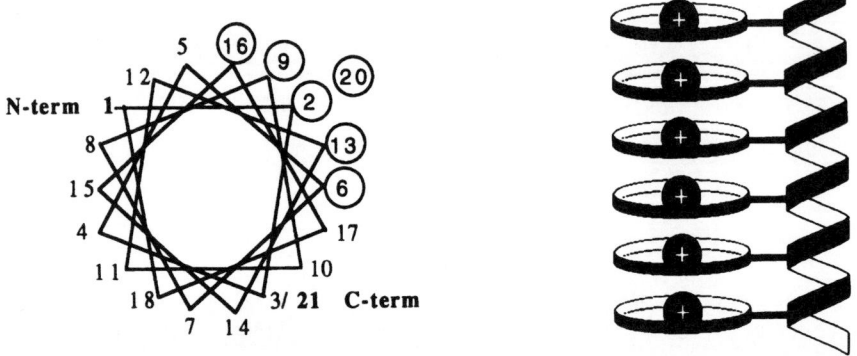

Fig. 24. *Left*: α-Helix axial projection of the artificial channel **69**. The position of the crown-ether residues is noted by circles. *Right*: proposed active form of **69** in a bilayer membrane. (Reproduced with the permission of Ref. 1)

predicted α-helix conformation and acts indeed as a functional ion channel in vesicles and in planar lipid bilayers. One advantage of our strategy is the possibility of easily synthesizing several analogs of **69** and therefore the ability to "engineer" its properties, such as the ionic selectivity and the lipophilicity. In addition, the segment condensation synthesis allows reasonable quantities of compounds such as **69** to be prepared and enables them to be further chemically modified for specific applications.

3.3 Novel Materials

Peptide nanostructures can be used in the preparation of "smart" new materials (a rapidly expanding area of research). The ability of polypeptides to adopt regular secondary structures in solution caught the attention of polymer chemists a long time ago. Several studies [58] have been done to characterize the structure–properties relationship in natural structural polypeptides, such as collagen [59], silk fibroin [60], and others. Also in the past 15 years, several groups used polypeptide frameworks to develop electron transfer devices [1, 61]. An interesting application is the preparation of photochromic materials for optical recording such as the poly-L-lysine derivative **70** [62]. It was shown that the CD spectrum of **70** could be modulated by the irradiation of the azobenzene side chains, which undergo a cis-trans isomerization.

Another important goal in this area is the preparation of well-defined materials and surfaces. Seminal work by Tirrell, Fournier, and co workers [63]

$$\text{H-(K-K-HN-}\overset{\displaystyle X}{\underset{\displaystyle H}{\text{C}}}\text{-CO)}_n\text{-OH} \quad \mathbf{70} \quad X= $$

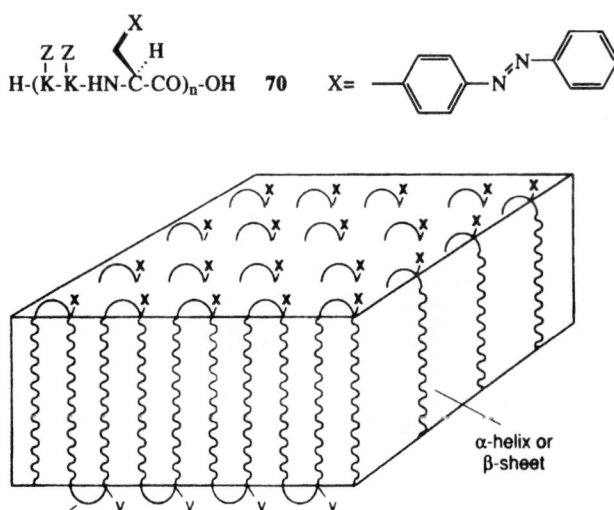

Fig. 25. The development of lamellar molecular "crystals" using polypeptides with a repeating sequence. The correct folding of these macromolecules produced by molecular biology gives peptide nanostructures, bearing specific functional groups in a well-defined spatial relationship. (Reproduced with the permission of Ref. 63)

$$\text{H-[(G-A)}_3\text{-G-HN-}\overset{\overset{\displaystyle X}{\mid}}{\underset{}{C}}\text{-CO]}_n\text{-OH}$$ **71** X= 〈benzene ring〉 or 〈thiophene ring〉 or —Se-CH$_3$

led to the construction of novel materials using a combination of chemical synthesis and molecular biology. As an example, polypeptides like **71** are designed to fold into lamellar molecular crystals that present a regular array of atoms or functional groups at their surfaces as shown schematically in Fig. 25 [64]. Another strategy to prepare well defined surfaces was reported recently by Whitesell and co workers [65]. The approach consists in synthesising α-helical peptides attached to a gold surface via a specific spacer using the N-carboxyanhydride method, as depicted in Fig. 26. Even though the polymerization affords

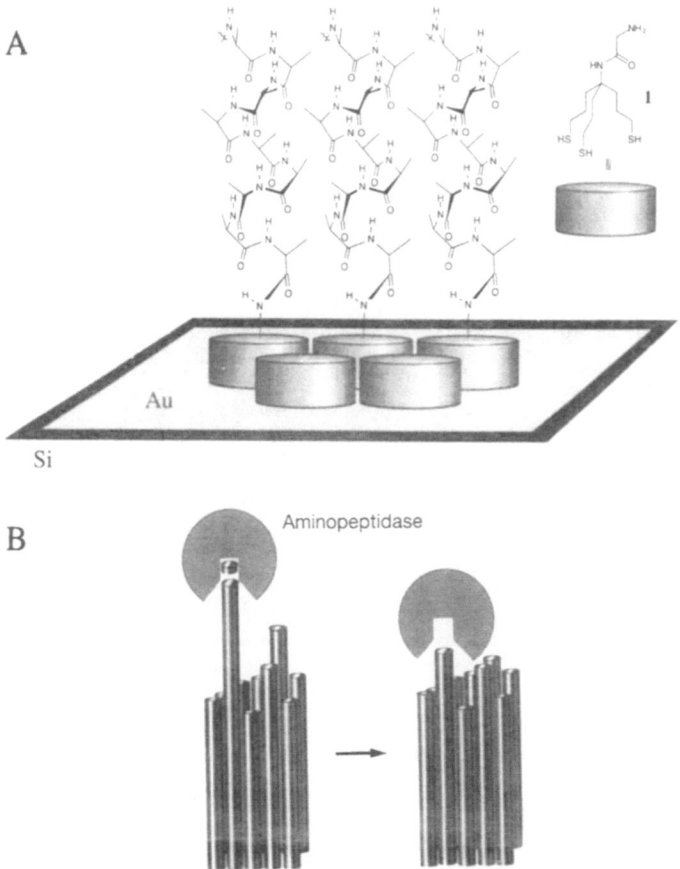

Fig. 26. The preparation of well-defined peptidic surfaces [65, 66]. **A** Using a thiol spacer, homopolypeptides (poly-A in this case) that adopt an α-helical conformation are synthesized on a gold surface. **B** Using an aminopeptidase, the longer peptidic chains are hydrolyzed to yield a more homogeneous surface. (Reproduced with the permission of Ref. 65, 66)

SH
|
7 2   Ac-NH-C-G-G-G-E-L-W-K-L-H-E-E-L-L-K-K-F-E-E-L-L-K-L-H-E-E-R-L-K-K-L-CONH$_2$

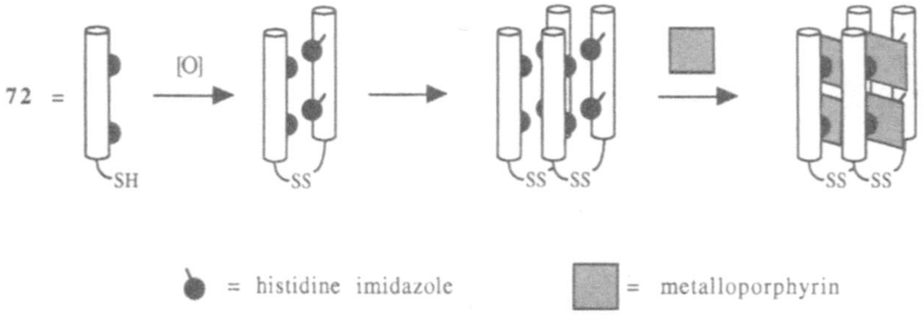

● = histidine imidazole   ■ = metalloporphyrin

Fig. 27. The preparation of a supramolecular peptide nanostructure by the selective complexation of four metalloporphyrins [67]. (Reproduced with the permission of Ref. 1)

peptide chains of different lengths (hence layers of different thicknesses), the authors have been able to level cleverly the surface using aminopeptidases that hydrolyze from the N-terminus the longer chain, producing a flatter surface (Fig. 26b) [66]. These new materials could find applications as waveguide optical switches.

3.4 Artificial Enzymes and Proteins

The development of synthetic enzymes and proteins has also been achieved through the preparation of structurally defined peptide nanostructures. A nice example, reported by DeGrado and co workers [67], is the construction (Fig. 27) of a four α-helix bundle system (**72**) that was shown to complex four metalloporphyrins by their axial coordination with the imidazole of the properly oriented histidines. This type of structure could be used as an artificial photosynthetic center. Along the same lines, Benson and co-workers [68] recently prepared a miniature hemoprotein, **73**, by linking two units of a 13-amino acid peptide to a porphyrin. UV-visible and CD studies confirmed that the metalloporphyrin is indeed sandwiched between the α-helical peptides, as depicted in **73**.

To design photo-induced electron transfer devices, the group of Nishino [69] reported the preparation of a bis-α-helical nanostructure, **74**. The peptidic framework was designed to orient rigidly in space the redox triad, composed of a Ru^{2+} complex, an anthraquinone, and two propylviologens, when incorporated in a lipid bilayer (Fig. 28). Although the compound exhibited a strong α-helical content in methanol, its conformation in a vesicle bilayer was different and undetermined. Nevertheless, irradiation of the Ru(II) complex of **74** resulted in a slow electron transfer.

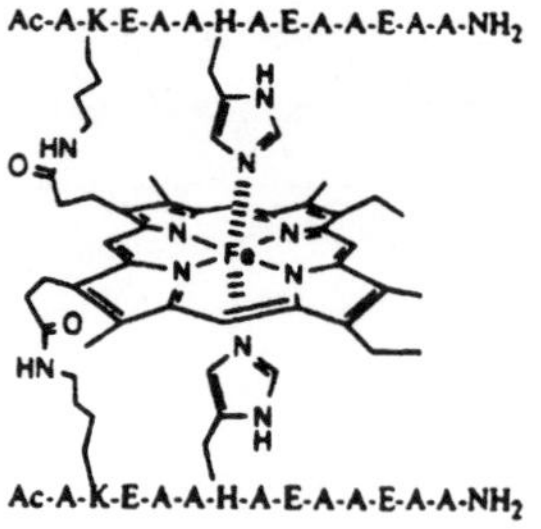

Ac-A-K-E-A-A-H-A-E-A-A-E-A-A-NH$_2$

73

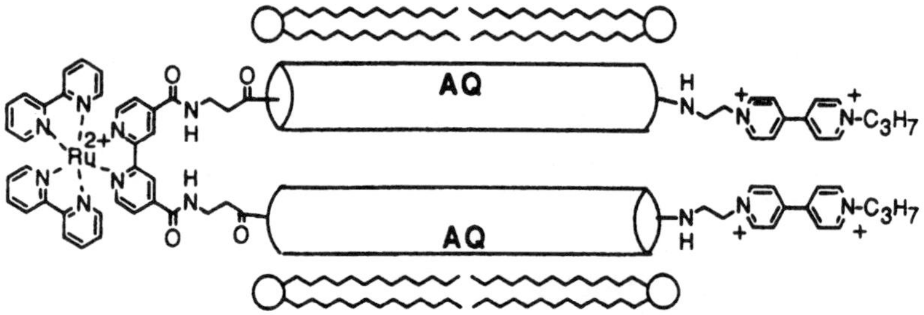

Ac-A-K-E-A-A-H-A-E-A-A-E-A-A-NH$_2$

74 Cylinders = -L-S-L-Aib-L-S-L-S-L-(AQ)-L-S-L-L-S-L-Aib-L-S-L-

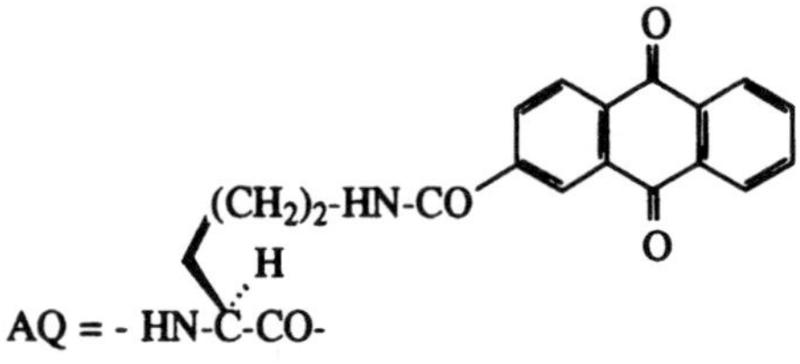

$$AQ = - HN-\overset{H}{\underset{(CH_2)_2-HN-CO}{C}}-CO-$$

Fig. 28. Schematic illustration of an artificial peptidic electron-transfer system designed to operate in a bilayer membrane. (Reproduced with the permission of Ref. 69)

From a totally different point of view, Morii and co-workers [70] have studied the utility of synthetic proteins as chiral hosts. They showed that α-helix bundle structures, formed by the folding of peptides such as **75**, induced chirality in fluorescent dyes by forming inclusion complexes in the hydrophobic interior of the structures. Again these results can have implications for the development of optical materials and switches.

The DNA binding ability of peptide nanostructures has also been reported by the artificial dimerization of peptide sequences corresponding to the contact region of the transcriptional activator protein GCN4. Cuenoud and Schepartz

75 $\left\{ \text{SCH}_2\text{-C-G-E-L-K}_2\text{-L}_2\text{-E}_2\text{-L-K}_2\text{-L}_2\text{-E}_2\text{-A-K-G-K-P-G}_2\text{-L-K}_2\text{-L-E}_2\text{-L}_2\text{-K}_2\text{-L-E}_2\text{-L}_2\text{-G-NH}_2 \right\}_2$
$\quad\quad\quad\; | $
$\quad\quad\quad\text{NH}_3{}^+$

[71] synthesized a 29-residue peptide linked to a terpyridyl group (**76**). Addition of Fe^{2+} yielded the dimer **77** which bound DNA more selectively (only at the CRE site) than GCN4 itself. In a similar fashion, the group of Morii [72] demonstrated that a 23-residue peptide from GCN4 (**78**), having a β-cyclodextrin or an adamantane at the N-terminal position, could be dimerized by the formation of an inclusion complex (**79**). This dimer bound DNA more tightly than the two peptide units themselves. These results illustrate again the potential of peptide nanostructures for therapeutic applications.

Finally, three groups have reported the preparation of artificial enzymes with catalytic activity. Stewart and co-workers [73] incorporated a "catalytic triad" from the serine proteases into a designed four α-helix protein (**80**). In their design, they incorporated one of the amino acids involved in the catalytic function at the N-terminal side of the α-helices that are linked together by their C-terminal position (Fig. 29). The authors proposed that the oxanion hole and the hydrophobic binding pocket are created by the three-dimensional structure formed by the folding of **80**. Compared to the spontaneous reaction, impressive

7 6 Ac-S-A$_2$-L-K-R-A-R-N-T-E-A$_2$-R$_2$-S-R-A-R-K-L-Q-R-M-K-Q-G-G-C-NH$_2$

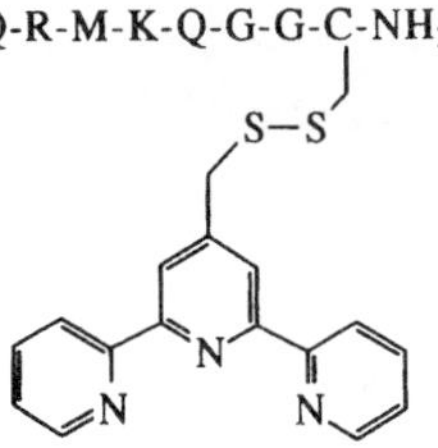

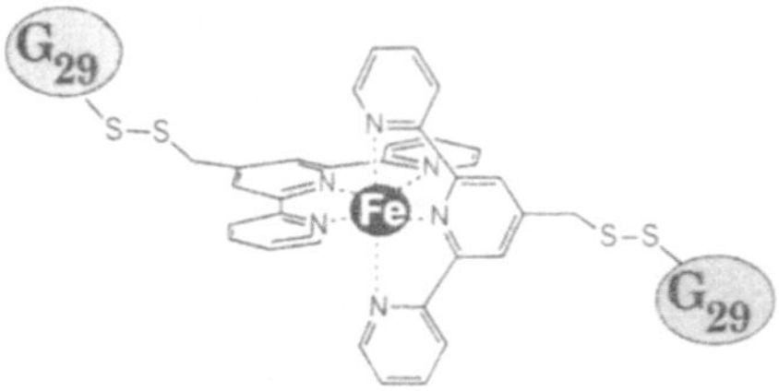

7 7

X-D-P-A$_2$-L-K-R-A-R-N-T-E-A$_2$-R$_2$-S-R-A-R-K-L-Q-C-NH$_2$

78 X= ß-cyclodextrin-CH$_2$- {G23CD} **or**

X= adamantane-CH$_2$-HN-CO-CH$_2$- {G23AD}

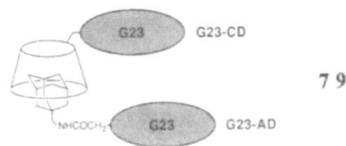

7 9

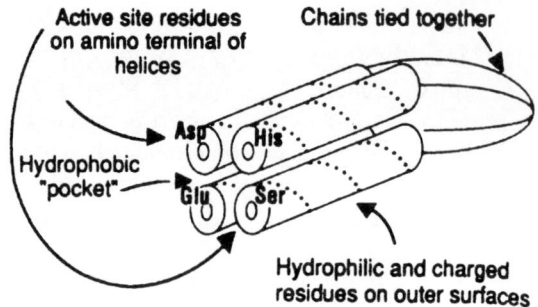

Fig. 29. Schematic representation of a synthetic protease [73]. Four α-helical peptides bearing a specific catalytic group are linked together. The correct folding of **80** brings the catalytic groups into a relationship that enables them to catalyze hydrolytic processes. (Reproduced with the permission of Ref. 73)

rate enhancements in the 10^5 range have been observed in the hydrolysis of acetyl-L-tyrosine ethyl ester, a known chymotrypsin substrate. They also showed that **80** has kinetic data typical of natural enzymes.

On the other hand, Sasaki and Kaiser [34a] have demonstrated that the already-mentioned artificial Fe^{3+} protein **39** has a hydroxylase activity with

Ac-E₂-A-E₂-K-A-K-R-L₂-E₂-L-K₂-A ─┐
  Ac-H-E₂-A-K₃-A-E-K-L₂-E₂-L-K₂-L-K ─┐
Ac-D-E-A-G-K₂-A-E₃-L-K₂-L₂-E₂-L-K₃ ── Orn-NH₂   **8 0**
  Ac-S-E-K-A-K₂-L₂-E₂-L-K₂-L-A ─┘

H_2 N-Leu-Ala-Lys-Leu-Leu-Lys-Ala-Leu-Ala-Lys-Leu-Lys-Lys-CONH $_2$

8 1

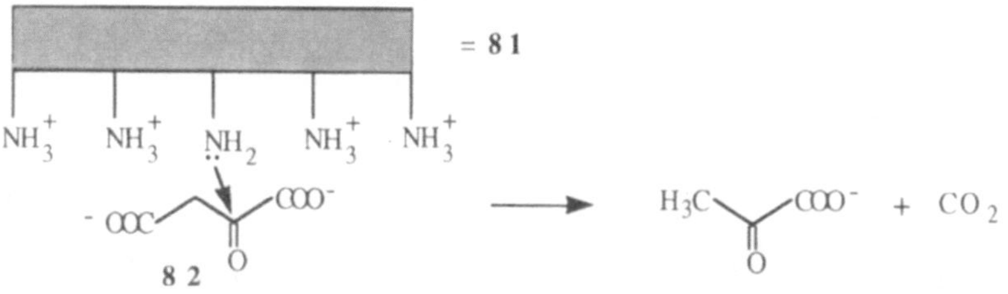

Fig. 30. The development of a synthetic decarboxylase (**81**) [74]. An α-helical peptide framework orients correctly several amino groups for the catalysis of the decarboxylation of **82**. (Reproduced with the permission of Ref. 1)

aniline similar to the one of a natural enzyme. They proposed that the catalytic activity of **39** is due to the binding of aniline in a hydrophobic pocket formed near the porphyrin by the four peptidic structures (Fig. 16).

Another functional artificial enzyme was developed by the group of Benner [74]. In a very simple approach, they prepared a short 14-residue peptide (**81**), designed to act as a decarboxylase, by orienting appropriately five amino groups of lysine side chains when adopting an α-helical conformation. The proper positioning of these groups allows **81** to catalyze 100-fold the decarboxylation of the substrate **82**, as shown in Fig. 30. Thorough structural and kinetic studies demonstrated that **81** adopts the predicted α-helical conformation and that the catalytic activity is imparted by the array of amino groups. Even though the rate enhancement is far from the one of the natural enzyme (10^8), this last example illustrates well the potential of structurally well-defined peptide structures, even as simple as **81**, for developing tailor made catalysts.

4 Conclusions and Prospects

As can be seen from the examples presented here, chemists have been quite successful in developing strategies for preparing structurally defined peptidic molecules in the nanometer range. In addition to shedding light on important biological phenomena such as protein folding, ligand induced conformational changes, and others, these fundamental studies constitute the groundwork of the peptide based nanotechnology. Although this field of research is still in its early stage of development, it is obvious that polypeptides are ideally suited for the synthesis of novel molecular materials with interesting and useful properties. These tailor made nanostructures will surely have an important impact on many areas of research in particular drug development, molecular electronics, and material science.

Acknowledgements. We are grateful to the NSERC of Canada, the Université de Sherbrooke, and the Fonds FCAR du Québec for the financial support of our work.

5 References

1. Voyer N, Lamothe J (1995) Tetrahedron 51: 9241
2. Schultz GE, Schirmer RH (1979) Principles of Protein Structure. Springer, Berlin Heidelberg New York
3. Creighton TE (1987) Nature 326: 547 and references cited therein
4. (a) Marquesee S, Baldwin RL (1987) Proc. Natl Acad Sci USA 84: 8898; (b) Padmanabhan S, Marquesee S, Ridgeway T, Laue TM, Baldwin RL (1990) Nature 344: 268; (c) Shoemaker KR, Kim PS, York EJ, Stewart JM, Baldwin RL (1987) Nature 326: 563

5. For excellent references on protein folding see: (a) Nall BT, Dill KA (eds) (1991) Conformations and Forces in Protein Folding, AAAS, Washington, DC; (b) Gierasch LM, King J (Eds) (1990) Protein Folding, AAAS, Washington, DC

6. (a) Ruan F, Chen Y, Hopkins PB (1990) J Am Chem Soc 112: 9403; (b) Ruan F, Chen Y, Itoh K, Sasaki T, Hopkins PB (1991) J Org Chem 56: 4347

7. (a) Ghadiri MR, Choi C (1990) J Am Chem Soc 112: 1630; (b) Ghadiri MR, Fernholtz, AK (1990) J Am Chem Soc 112: 9633

8. Dado GP, Gellman SH (1993) J Am Chem Soc 115: 12609

9. Mutter M, Hersperger R (1990) Angew Chem Int Ed Engl 29: 185

10. (a) Voyer N, Roby J, Deschênes D, Bernier J (1995) Supramol Chem 5: 61; (b) Voyer N, Roby J (1991) Tetrahedron Lett 32: 331; (c) Voyer N, Deschênes D, Bernier J, Roby J (1992) J Chem Soc Chem Commun 1992: 134

11. Voyer N, Guérin B (1994) Tetrahedron 50: 989

12. Imperiali B, Fisher SL (1991) J Am Chem Soc 113: 8527

13. Imperiali B, Kapoor TM (1993) Tetrahedron 49: 3501

14. (a) Rose GD, Gierasch LM, Smith JA (1985) Adv Protein Chem 37: 1; (b) Stradley SJ, Rizo J, Bruch MD, Stroup AN, Gierasch LM (1990) Biopolymers 29: 263

15. Ösapay G, Taylor JW (1992) J Am Chem Soc 114: 6966

16. Bracken C, Gulyás J, Taylor JW, Baum J (1994) J Am Chem Soc 116: 6966

17. Jackson DY, King DS, Chmielewski J, Singh S, Schultz PG (1991) J Am Chem Soc 113: 9391

18. Lutgring R, Chmielewski J (1994) J Am Chem Soc 116: 6451

19. Futaki S, Kitigawa, K (1994) Tetrahedron Lett 35: 1267

20. (a) Presta LG, Rose GD (1988) Science 240: 1632; (b) Harper ET, Rose GD (1993) Biochemistry 32: 7605

21. Richardson JS, Richardson DC (1988) Science 240: 1648

22. Forood B, Reddy HK, Nambiar KP (1994) J Am Chem Soc 116: 6935

23. During the preparation of this chapter, a review on the use of templates to induce peptide secondary structures appeared: Scheiner JP, Kelly JW (1995) Chem Rev 95: 2169

24. (a) Kemp DS, Boyd JG, Muendel CC (1991) Nature 352: 451; (b) Kemp DS, Curran TP, Davis WM, Boyd JG, Muendel C (1991) J Org Chem 56: 6672; (c) McClure KF, Renold P, Kemp DS (1995) J Org Chem 60: 454 and references cited therein

25. (a) Kemp DS, Bowen BR (1988) Tetrahedron Lett 29: 5077; (b) ibid 29: 5081

26. (a) Feigel M (1986) J Am Chem Soc 108: 181; (b) Wagner G, Feigel M (1993) Tetrahedron 49: 10831; (c) Brandmeier V, Sauer WHB, Feigel M (1994) Helv Chim Acta 77: 70

27. (a) Díaz H, Espina JR, Kelly JW (1992) J Am Chem Soc 114: 8316; (b) Díaz H, Tsang KY, Choo D, Espina JR, Kelly JW (1993) J Am Chem Soc 115: 3790; (c) Díaz H, Tsang KY, Choo D, Kelly JW (1993) J Am Chem Soc 115: 3790

28. Schneider JP, Kelly JW (1995) J Am Chem Soc 117: 2533

29. Nowick JS, Abdi M, Bellamo KA, Love JA, Martinez EJ, Noronha G, Smith ER, Ziller JW (1995) J Am Chem Soc 117: 89

30. Nowick JS, Lee IQ, Mackin G, Pairish M, Shaka AJ, Smith ER, Ziller JW (1996). To appear in Wilcox CS, Hamilton AD (eds) Bioorganic Catalysis. Kluwer, Dordrecht th. The author thanks J. Nowick for providing a preprint of this publication

31. Mutter M, Vuillemier S (1989) Angew Chem Int Ed Engl 28: 535

32. Mutter M, Tuchscherer GG, Miller C, Altmann K-H, Carey RI, Labhardt AM, Rivier JE (1992) J Am Chem Soc 114: 1463

33. Montal M, Montal MS, Tomich JM (1990) Proc Natl Acad Sci USA 87: 6929

34. (a) Sasaki T, Kaiser ET (1989) J Am Chem Soc 111: 380; (b) ibid (1990) Biopolymers 29: 79

35. Åkerfeld KS, Kim RM, Camac D, Groves JT, Lear JD, DeGrado WF (1992) J Am Chem Soc 114: 9656

36. Mihara H, Nishino N, Hasegawa R, Fujimoto T (1992) Chem Lett 1992: 1805

37. Mihara H, Nishino N, Fujimoto T (1992) Chem Lett 1992: 1809

38. Åkerfeld KS, DeGrado WF (1994) Tetrahedron Lett 35: 4489

39. Gibb BC, Mezo AR, Sherman JC (1995) Tetrahedron Lett 36: 7587

40. (a) Lieberman M, Sasaki T (1991) J Am Chem Soc 113: 1470; (b) ibid (1993) Tetrahedron 49: 3677; (c) Lieberman M, Tabet M, Sasaki T (1994) J Am Chem Soc 116: 5053

41. Ghadiri MR, Soares C, Choi C (1992) J Am Chem Soc 114: 825

42. Ghadiri MR, Case MA (1993) Angew Chem Int Ed Engl 32: 1594

43. Ghadiri MR, Soares C, Choi C (1992) J Am Chem Soc 114: 4000

44. Minoura N (1993) J Chem Soc Chem Commun 1993: 196
45. Berthet M, Yordanov S, Sonveaux E (1986) Makromol Chem Rapid Commun 7: 205
46. Berthet M, Sonveaux E (1986) Biopolymers 25: 189
47. (a) Anzai J, Ueno A, Osa T (1982) Makromol Chem Rapid Commun 3: 55; (b) Anzai J, Ueno A, Osa T (1980) Makromol Chem Rapid Commun 1: 741
48. Voyer N, Maltais F (1993) Adv Mater 5: 568
49. Voyer N, Guérin B (1992) J Chem Soc Chem Commun 1992: 1253
50. Hibert MF, Hoflack S, Truwpp-Kallmeyer S, Bruinvels A (1993) Médecine/Sciences 9: 31
51. LaBrenz SR, Kelly JW (1995) J Am Chem Soc 117: 1655
52. Hille B (1984) Ionic Channels of Excitable Membranes, Sinauer, Sunderland, MA, USA
53. Voyer N, Robitaille M (1995) J Am Chem Soc 117: 6599 and references cited therein
54. Åkerfeldt KS, Lear JD, Wasserman ZR, Chung LA, DeGrado WF (1993) Acc Chem Res 26: 191
55. Ghadiri MR, Granja JR, Buehler LK (1994) Nature 369: 301
56. Granja JR, Ghadiri MR (1994) J Am Chem Soc 116: 10785
57. Voyer N (1991) J Am Chem Soc 113: 1818
58. (a) Elliott A (1987) In: Fibrous Protein Structure. Academic Press, New York, p 117; (b) Walton AG (1981) Polypeptides and Protein Structure. Elsevier, New York
59. van der Rest M, Bruckner P (1993) Curr Opin Struc Biol 3: 430
60. Voet D, Voet JG (1995) Biochemistry. Wiley, New York, p 153
61. Sisido M (1992) Prog. Polym. Sci. 17: 699
62. Sisido M, Ishikawa Y, Itoh K, Tazuke S (1991) Macromolecules 24: 3993 and 3999
63. (a) Tirrell JG, Fournier MJ, Mason TL, Tirrell DA (1994) Chem Eng News December 19: 40; (b) Stinson SC (1990) Chem Eng News July 16: 26; (c) Tirrell DA, Fournier MJ, Mason TL (1991) MRS Bulletin 16: 23
64. Kothakota S, Mason TL, Tirrell DA, Fournier MJ (1995) J Am Chem Soc 117: 536
65. Whitesell JK, Chang HK (1993) Science 261: 73
66. Whitesell JK, Chang HK, Whitesell CS (1994) Angew Chem Int Ed Engl 33: 871
67. Robertson DE, Farid RS, Moser CC, Urbauer JL, Mulholland SE, Pidikiti R, Lear JD, Wand AJ, DeGrado WF, Dutton PL (1994) Nature 368: 425
68. Benson DR, Hart BR, Zhu X, Doughty MB (1995) J Am Chem Soc 117: 8502
69. Mihara H, Nishino N, Hasegawa R, Fujimoto T, Usui S, Ishida H, Ohkubo K (1992) Chem Lett 1992: 1813
70. (a) Morii H, Ichimura K, Uedaira H (1990) Chem Lett 1990: 1987; (b) ibid (1991) Proteins Struc Funct Gen 11: 133
71. Cuenoud B, Schepartz A (1993) Science 259: 510
72. Ueno M, Murakami A, Makino K, Morii T (1993) J Am Chem Soc 115: 12575
73. Hahn KW, Klis WA, Stewart JM (1990) Science 248: 1544
74. (a) Johnsson K, Alleman RK, Widmer H, Benner SA (1993) Nature 365: 530; (b) Johnsson K, Alleman RK, Benner SA (1990) In: Bleasdale C, Golding BT (eds) Molecular Mechanisms in Bioorganic Processes, RSC, Cambridge, UK, p 166

Cytochrome P450: Significance, Reaction Mechanisms and Active Site Analogues

Wolf-Dietrich Woggon

Institut Für Organische Chemie der Universität Basel St. Johanns-Ring 19
CH-4056 Basel

Topics in Current Chemistry, Vol. 184
© Springer-Verlag Berlin Heidelberg 1996

1 Mechanisms of Cytochrome P450 Catalyzed Oxidations

1.1 Introduction

The significance of cytochrome P450 catalyzed reactions has now been recognized for more than 30 years [1] as one of nature's most common and sophisticated methods to oxidize endogeneous compounds in procaryotic and eucaryotic organisms. Accordingly, P450 enzymes can be isolated from bacteria [2] plants [3] and different tissues of mammals [4], where numerous isozymes participate highly specifically in the metabolism of compounds as diverse as steroids, fatty acids and alkaloids. These enzymes display a surprisingly large repertoire of reactions [5] such as the (multiple) C-hydroxylation of aliphatic and aromatic compounds, epoxidations, and the dealkylation at oxygen, nitrogen and sulfur. The same reactivity is also common to those cytochromes P450 acting on xenobiotics, which can be isolated, e.g., from liver microsomes. The hepatic P450s are known to be induced by exogeneous compounds and generally accept a broad spectrum of substrates, which is understood as a method of detoxification in order to render lipophilic compounds water soluble and excretable.

The inherent reactivity of these enzymes is attributed to an iron protoporphyrin IX complex which is bound to the protein via hydrogen bridges of the two propionate side chains and through a thiolate ligand coordinating to the iron (Fig. 1). The latter is delivered by a cysteine residue in a highly conserved area of the protein, and is placed at the face of the porphyrin opposite to the binding site of oxygen and the substrate.

The presence of the Cys-S$^-$ ligand became obvious from early spectroscopic investigations of crude P450 preparations which revealed a considerable red shift of the *Soret-band* from 420 nm to 450 nm on treatment of the reduced form of P450 with carbon monoxide [1, 6]. This finding has been supported by many studies with synthetic active site models [7, 8] as well as by theoretical calculations indicating that the degenerate *Soret-band* of the Fe(II)-S$^-$ porphyrinato complex must split into two bands of equal intensity at ~ 380 nm and ~ 450 nm on the addition of carbon monoxide [9]. The formation of this OC-Fe(II)-S$^-$ complex has been used from the early beginning for identifying P450 enzymes in living tissues and for proving the significance of synthetic iron porphyrins as enzyme models. It was suggested that the electron donating character of the proximal thiolate ligand fine-tunes the redox potential of the system and triggers the unique chemical P-450 reactivity, in particular the reductive cleavage of molecular oxygen and the "O" insertion into non activated C–H bonds [10, 11], see below. Substrate recognition and the stereo- and regiospecificity of "O"-insertion is dependent on certain distal amino acid residues at the binding domain.

Amino acid sequences were determined for about 70 individual cytochromes P450, although X-ray structures have been obtained from only three

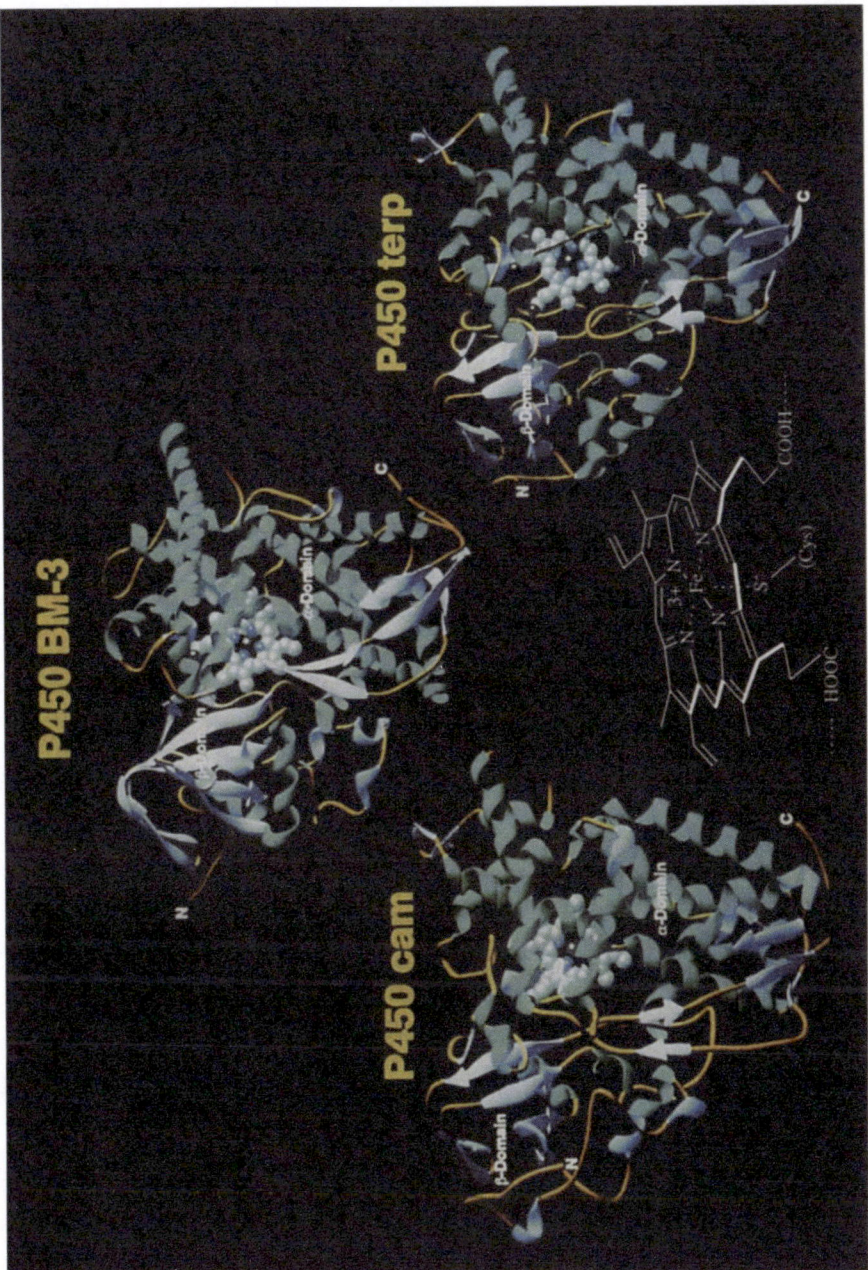

Fig. 1. X-ray structures of cytochrome P450$_{cam}$, P450 BM-3 and P450 terp; underneath iron(III) protoporphyrin IX bound to the protein via hydrogen bridges (...) and the thiolate ligand. Printed with permission of J.A. Peterson [14].

non-membrane bound bacterial proteins: $P450_{BM-3}$, [12], $P450_{cam}$ [13] and $P450_{terp}$ [14], Fig. 1. Like other class I enzymes, the latter two proteins, isolated from *Pseudomonas* strains, require an electron transfer system comprising a FAD containing reductase and an iron-sulfur protein similar to eukaryotic P450s associated with mitochondrial membranes. In contrast $P450_{BM-3}$ from *Bacillus megaterium* interacts directly with a covalently bound reductase containing FAD and FMN. Accordingly BM-3 is termed a Class II P450, and viewed as a prototype for microsomal P450s. For the inexperienced eye the overall folding of the three proteins looks rather similar, but the relative arrangements of the α-helices and β-sheets differ substantially – see Fig. 1. Furthermore, the amino acid sequence identity between them is only 10%.

In view of the different substrates hydroxylated by these enzymes, a closer look at the binding site is of particular interest. For BM-3, access to the heme is provided by a long hydrophobic channel, 8–10 Å in diameter, which was identified as the binding site for the long chain fatty acid substrates [12]. Close to the protein surface the binding pocket is "controlled" by a flexible Arg47 residue possibly adjusting the channel for the hydroxylation of fatty acids with various chain lengths. At the other end of the pocket, Phe87 seems to be of significance for substrate orientation since its phenyl ring is directed perpendicular to the porphyrin and appears suitable for hydrophobic interactions with the substrate.

The results concerning the structure of BM-3 are very recent and important due to its functional analogy and its sequence similarity to the microsomal P450s. The latter aspect was always considered to be a problem concerning $P450_{cam}$ because it was questioned whether data of this structurally distinct protein are useful as a general model for eukaryotic P450s. It now looks like this caution in extrapolating $P450_{cam}$ results is justified with respect to the oxygen/substrate binding domain and the attachment of the reductase. Nevertheless, the comprehensive structural information about different forms of $P450_{cam}$ makes it a unique example to discuss P450 catalysis.

Using camphor **1** as the only carbon source, *Pseudomonas putida* employs $P450_{cam}$ in order to catalyze the stereospecific hydroxylation at the 5-*exo* position **1** $\Rightarrow$ **2** as a first step in a cascade of energy supplying reactions (Fig. 2).

The information gained from the X-ray structures of the substrate-free heme protein [15], the E·S complexes with camphor [13] and substrate analogues [16], and the structure of the CO adduct [17] of the reduced form of $P450_{cam}$ had a great impact on P450 research. The comparison of these "derivatives" of $P450_{cam}$ with respect to changes of the spatial arrangements at the active site not only confirmed former "indirect" spectroscopic evidence but provided important details of substrate binding **3** $\Rightarrow$ **4**, and clues concerning the mechanism of P450 action. Thus, they define the starting point of the catalytic cycle as depicted in Fig. 3.

In the absence of camphor the substrate binding site of the resting state **3** is occupied by a cluster of six hydrogen bonded water molecules. Surprisingly the iron(III) of the resting state **3** is predominantly $\sim 93\%$ low spin ($S = \frac{1}{2}$) [18].

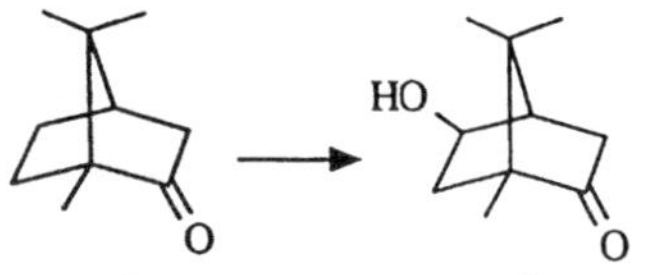

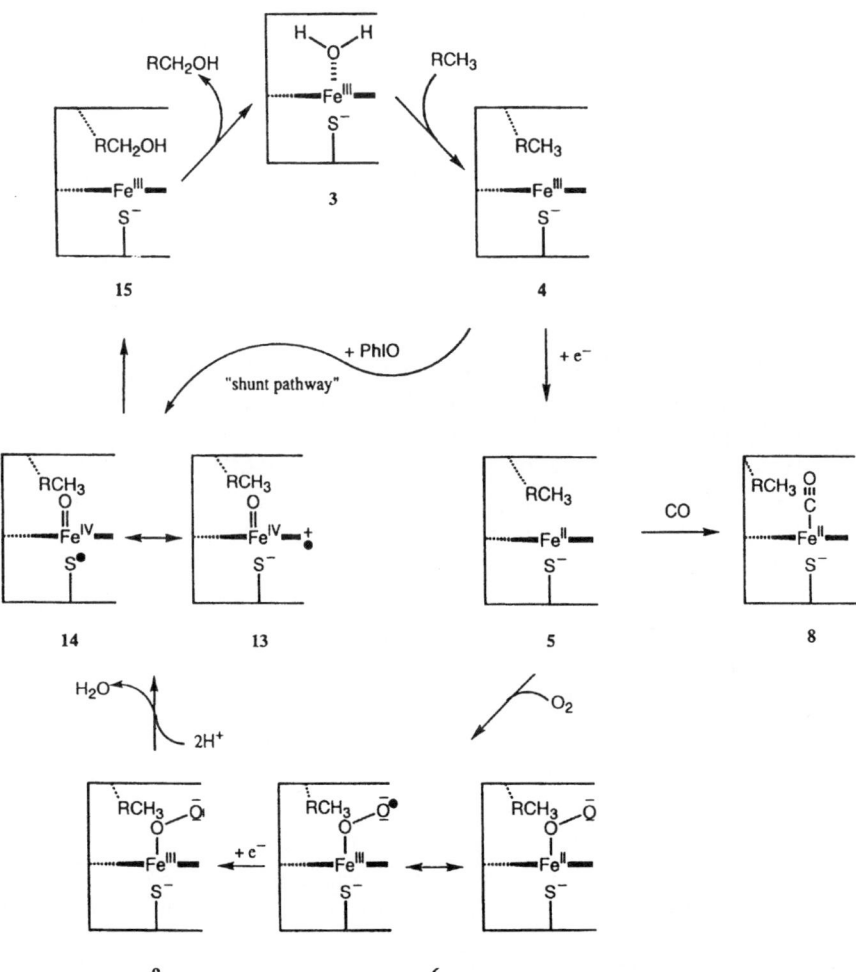

Fig. 2. Hydroxylation of camphor **1** by cytochrome P450$_{cam}$

Fig. 3. Catalytic cycle of cytochrome P450

This is difficult to understand since the presence of water as a sixth ligand is insufficient to explain the low spin character of **3**. Two arguments have been put forward recently, one assuming the coordination of an HO$^-$ embedded in the H$_2$O-network [19], the other identifying a significant role for the electrostatic field of the protein stabilizing the low-spin state in the presence of the water cluster [20]. On binding of camphor the water cluster is completely removed, including the sixth ligand, and the spin equilibrium is shifted to $\sim 97\%$ high spin iron(III) $(S = \frac{5}{2})$, shortening slightly the distance of the iron to S^- ligand [21]. Simultaneously, the redox potential changes [21] from -300 mV (3) to -170 mV (4) [22] in order to facilitate the subsequent reduction $4 \Rightarrow 5$ by one electron delivered from NADPH via the redox protein putidaredoxin $(-196$ mV). Binding of O_2 seems to be kinetically different in model systems as compared to native enzymes but eventually affords a compound formally described as a diamagnetic, low spin iron(III) complex 6.

Two models for the latter complex are known. The synthetic analogue 7 consists of an assembly of iron(II)*meso*-tetrakis ($\alpha,\alpha,\alpha,\alpha$-o-pivalamidophenyl)-porphyrin, cryptated sodium tetrafluorothiophenolate and molecular oxygen [23]. The Mössbauer spectra and the crystal structure of the adduct confirm the bent, end on binding of O_2 to the iron inside the *picket-fence* cavity and indicate a certain disorder of the distal oxygen atom having a dynamic distribution over three different sites due to non bonded interactions with the pivalylamido groups. The axial thiolate ligand is coordinating to the iron from the unprotected face of the porphyrin (Fig. 4). The binding of O_2 to iron, and the interaction of the terminal oxygen with pivalyl groups has been also investigated by ^{17}O-NMR spectroscopy [24].

The comparison of the X-ray structures of the iron(III)-camphor P450$_{cam}$ 4 [13] and the ternary complex OC-iron(II)-camphor-P450$_{cam}$ 8 [17] provides important information on geometrical changes at the substrate binding domain likely to be relevant to the formation of the corresponding O_2-complex 6 [19]. Binding of CO to 5 gives a low spin complex; the iron moves into the plane of the macrocycle and the Fe-S$^-$ distance is stretched by 0.21 Å. CO binds slightly distorted from the heme normal and forces the substrate to move away from the heme by 0.8 Å towards the substrate access channel. Accordingly camphor is

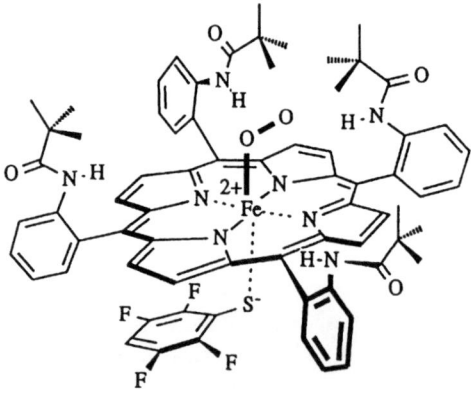

7

Fig. 4. Structure of 7, a model compound for the O_2 complex of cytochrome P450 6

gaining more mobility though the hydrogen bond between the carbonyl group of **1** and Tyr96 remains intact. Furthermore, CO-binding induces a slight distortion of the distal helix suggesting a putative O_2-binding site in this area. It seems likely that in the process of reductive O_2-bond scission all these spatial changes may be at least in part reversed towards the situation of the iron(III)-camphor P450$_{cam}$ directing camphor into the position to become hydroxylated only at C(5).

In order to facilitate the reductive cleavage of oxygen a second electron is added to **6** by P450 reductase or cytochrome b_5 to yield **9**; several experiments indicate that this reduction is the rate limiting step of the catalytic cycle [25]. It was suggested that acylation of the peroxo group supports heterolytic O_2 cleavage [26].

Subsequent model studies [27] seemed to confirm this idea, demonstrating that the high spin acylperoxo iron(III) porphyrin **10** rearranged in a fast, acid catalyzed, first order reaction to an oxo-iron(IV) porphyrin radical cation **11**, Fig. 5. A large negative entropy term was estimated and explained by coordination of **10** with an unidentified ligand L to form a six coordinate, low spin state **12**. The spectroscopically well defined **11** displayed chemical reactivity characteristic of native P450, and it seems to be equivalent to the species **13** one can generate on treating **4** with artificial "O" donors like iodosobenzene – see "shunt-pathway" Fig. 3. However the formation of **11** via an acylperoxo iron(III) porphyrin has only limited significance to the enzymatic reaction, because the X-ray of cytochrome P450$_{cam}$ [13] revealed no acylating residue at the protein near the oxygen binding site. Moreover, this investigation, like many other discussions concerning the catalytic cycle of P450, disregards the presence of the thiolate ligand coordinating to the iron.

In this context it is noteworthy that a suitable synthetic doubly bridged iron porphyrin carrying a thiolate ligand performs oxygen cleavage in the absence of an acylating agent [28], and almost certainly by homolytic O–O bond scission – see below. Homolysis of the O–O bond is also discussed in order to explain the surprisingly nonspecific H· abstraction in the course of the overall stereospecific hydroxylation of camphor in the 5-*exo*-position [29]– see below.

In contrast, investigating the reaction of "chelated protohemin chloride" with H_2O_2, peracids, and *tert*-BuOOH, respectively, in buffered solution, suggested that general acid catalysis by water in the active site may facilitate the

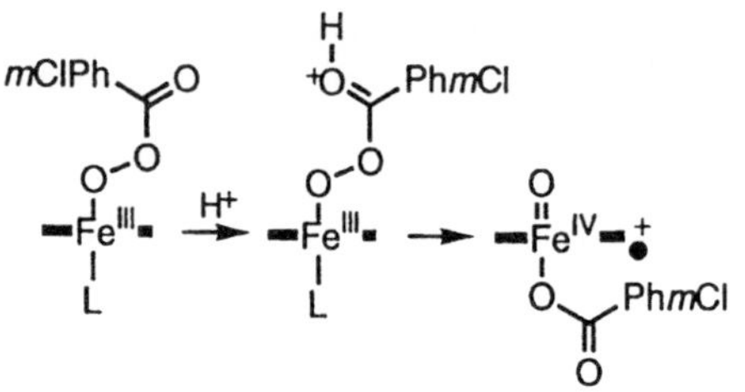

Fig. 5. Preparation of an iron(IV)oxo porphyrin radical cation; *bold lines.* refer to *meso*-tetra mesityl porphine

heterolytic O–O bond cleavage [30]. However, this interpretation contradicts the results obtained from the X-ray structures of **4** and **8** as discussed above, which revealed the absence of water when the substrate is bound.

Recently site directed mutagenesis of polar amino acids near the oxygen binding domain revealed the significance of Thr252 and Asp251 [31]. For example, the Thr252Ala mutant displayed rapid autoxidation and little substrate hydroxylation, whereas a drastic decrease in catalytic turnover was observed when Asp251 was replaced by Asn.

The kinetic and thermodynamic parameters determined for a number of amino acid modifications indicated that the OH group of Thr252 serves to stabilize both the one- and two-electron reduced oxygen intermediates **6** and **9**, Fig. 3. Moreover, assuming a certain conformational mobility in this area, the COOH group of Asp251 appears to be hydrogen bonded to the OH group of Thr252 and to the amino functions of the solvent accessible Asp182-, Lys178-, Arg186-network. Accordingly, it was suggested that Asp251 serves as a proton shuttle between these amino acids, and Thr252 provides a mechanism of O–O bond cleavage by means of general acid catalysis, and assists generating the high valent iron-oxo intermediate [32].

It is generally believed that porphyrin radical cations like **13** and **11** are electronic representations of the oxygen transfer intermediate of the P450 cycle. However, in view of the proper orientation of the thiolate orbitals to the d_{yz}/d_{xz} orbitals of iron and the pronounced spin orbit coupling of sulfur, it seems likely that the S^- ligand from cysteinate acts as an electron donor supporting O–O cleavage. Accordingly, **14** seems to be significant resonance form of the oxo-iron intermediate reflecting the structural and electronic parameters of the coordination sphere of iron. All evidence available indicates that it is an electron deficient species such as **14↔13** that performs regio- and stereospecific O-insertion by a radical abstraction recombination mechanism [33] to give **15**, from which the product $R\text{-}CH_2\text{-}OH$, e.g. **2**, is released in order to reestablish the resting state **3** see Fig. 3.

It becomes clear from the brief discussion of the catalytic cycle of cytochrome P450 that solid information is only available on the structures of **3** and **4**. The subsequent reactions and remaining intermediates have been characterized or proposed by investigating the mechanisms of P450 action with suitable enzymes or synthetic model compounds. In this context the preparation of active site analogues is of significance in as much as they can be regarded as spectroscopic equivalents of "missing links" and furthermore display the inherent reactivity characteristic of cytochrome P450. The latter aspect implies the possible application of P450 models in metabolism research and in synthesis.

1.2 Hydroxylations of Nonactivated C–H Bonds

Only very few reagents are available to the organic chemist which are applicable to the hydroxylation or oxidation of nonactivated C–H bonds, namely O_3/SiO_2

[34], dioxiranes [35], perfluorodialkyloxaziridines [36], and various iron pyridine complexes, the so called "Gif chemistry" [37]. All but the last of these reactions display regioselectivities which parallel those of radical reactions, thus *tert*-C–H bonds are preferentially attacked over *sec*- and primary positions. However, the regio- and stereospecific O-insertion into nonactivated C–H bonds catalyzed by cytochromes P450 is without precedent in nonenzymatic chemistry. The most striking example is the hydroxylation of alkanes which, due to the lack of a functional group, have to be oriented in the active site only by hydrophobic interactions with the protein [12]. In this context results concerning the hydroxylation of *n*-octane are briefly discussed [38]. After incubation of $(1R)(1-^3H,^2H)$-*n*-octane **16** with P450-containing rat liver microsomes, a mixture of octanols **17–19** was isolated. As determined by equilibration with HLAD/NAD/porcine heart diaphorase, the major product of the octanols, chiral at C(1), had the (*S*) configuration. Since complementary results were obtained with (*1S*)-*n*-octane it is safe to conclude that the hydroxylation of *n*-octane, chiral at C-1 due to isotopic labelling, proceeds with retention of configuration (Fig. 6).

The product composition was different when *n*-octane **20** was incubated with three different purified cytochromes P450, namely $P450_b$, $P450_c$, and $P450_{LM2}$ [39, 40]. In all cases a mixture of octanols **21–23** was obtained displaying a clear preference (74.2%) for hydroxylation at C(2) (Fig. 6). Using the *n*-octane **24**, trideuterated at one of the terminal methyl groups, and analyzing the composition of octanols **25–27**, significant intramolecular isotope effects were determined which were identical for all three enzymes within experimental error, $k_H/k_D = 11.8$. This value corresponds to the calculated intrinsic isotope effect D_K which is defined as the full effect from a single isotopically sensitive step of catalysis without interference of other isotopically insensitive steps [41]. These other rate factors usually lead to much smaller (masked) observed isotope effects. Accordingly the observed k_H/k_D is fully expressed in the P450 catalyzed hydroxylation of **24**.

Both studies provide important clues concerning the transition state geometry, because the observation of a fully expressed primary isotope effect indicates a rate determining H· removal by the oxo-iron intermediate **13**↔**14** in a highly symmetrical fashion. Furthermore, the observation of overall retention of configuration suggests a very fast trapping of the intermediate alkyl radical by the hydroxylated iron(IV)porphyrin in the "cage" of the active site.

The involvement of radical intermediates in P450 catalyzed reactions has been convincingly demonstrated by experiments with strained hydrocarbons, for which, after H· removal, radical pair recombination competes with rearrangement. In view of the fact that cyclopropylmethyl radicals rearrange to 3-butenyl radicals with a rate constant of $1 \times 10^8 s^{-1}$, the detection of rearranged products would be taken as evidence for radical formation and furthermore provides an estimate of the radical recombination rate. Since the P450 catalyzed hydroxylation of nortricyclane **28** gave only **29**, the product of immediate radical recombination, it was concluded that the latter process is faster than $10^8 s^{-1}$ (Fig. 7) [42, 43]. The rearrangement of radicals derived from

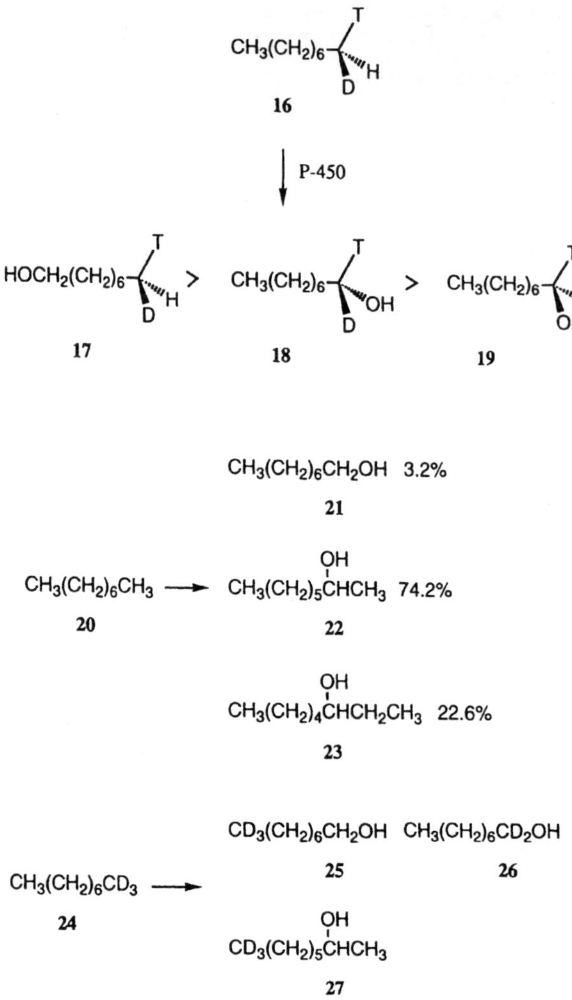

Fig. 6. Experiments concerning the P450-catalyzed hydroxylation of *n*-octane

bicyclo[2.1.0]pentanes was estimated to be ten times faster, and thus **30** was choosen as a suitable substrate for P450. After incubation of **30** with rat liver microsomes, the rearranged alcohol **31** and the product of fast radical recombination **32** were isolated in ratios 1 : 6 to 1 : 10 (Fig. 7). Moreover, it was shown by incubation of *exo-* and *endo*-deuterated **30** that the *endo*-alcohol **32** was produced under retention of configuration. The results confirm that the enzymatic hydroxylation is a stereospecific, nonconcerted process involving a "free" radical intermediate **33** that is trapped by the hydroxy-iron(IV) with $k > 10^{10}\,\mathrm{s}^{-1}$ [44].

Mechanistic studies of the hydroxylation of camphor have been supported by detailed information gained from crystal structures of several forms of cytochrome P450$_{cam}$. In order to hydroxylate specifically the 5-*exo* position,

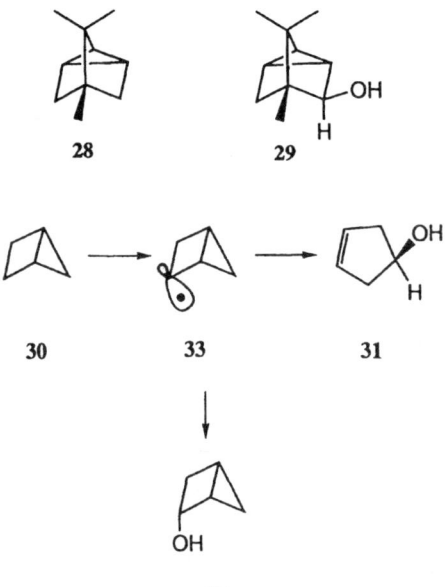

Fig. 7. Strained hydrocarbons as "radical clocks" to estimate the rate of radical recombination

the camphor molecule is bound ~ 4 Å above the porphyrin by three interactions [13]: a hydrogen bridge from the hydroxyl group of Tyr96 to the carbonyl, hydrophobic interactions between the substrate 8,9-*gem*-dimethyl groups and Val295, and between the camphor 10-methyl group and the Leu244-Val247 area of the protein, respectively (Fig. 8). The significance of these active site residues to camphor recognition and the stereospecificity of hydroxylation is also indicated by site directed mutagenesis of Val295, Val247, and Tyr96 [45]. From the X-ray of the OC-iron(II)-camphor P450$_{cam}$ [17] it was inferred that O-insertion into the nonactivated 5-*exo* C–H bond of **1** specifically occurs because the 5-*exo* hydrogen is about 1.5 Å closer than the 5-*endo* hydrogen to the carbon of CO which is likely to be the position of the oxygen in the oxo-iron(IV) intermediate **13**. Although the location of camphor in the active site after O–O bond scission is unknown, the "distance argument" seems to be useful in order to explain otherwise conflicting results obtained from incubation of 5-*exo*- and 5-*endo*-deuterated camphor **34** and **35** with P450$_{cam}$. In fact both substrates gave in a different ratio deuterated and undeuterated 5-*exo*-hydroxycamphor **36** and **37**, indicating that H· removal and O-insertion are separate events in the reaction sequence (Fig. 8) [46].

The product ratios **36**/**37** obtained from incubations of **34**, and **35** (Fig. 8) are reasonably well explained assuming a rather "masked" intramolecular isotope effect $(k_H/k_D)_{exo} \sim (k_H/k_D)_{endo} = 2$, and a twofold preference for the abstraction of *exo*-H· over *endo*-H·. This observation suggests that the oxo-iron(IV) intermediate assumes a position, different from the CO complex, dissecting the H-C(5)-H angle unsymmetrically, being slightly closer to the *exo*-hydrogen. For this

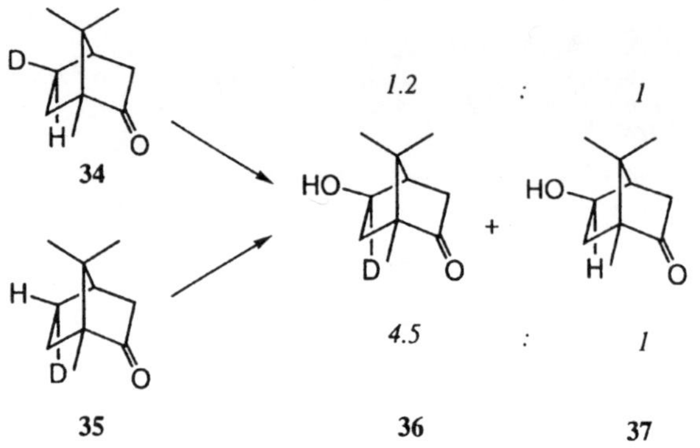

D
H
34
H
D
35
1.2 : 1
HO
D
O
+
HO
H
O
4.5 : 1
36
37

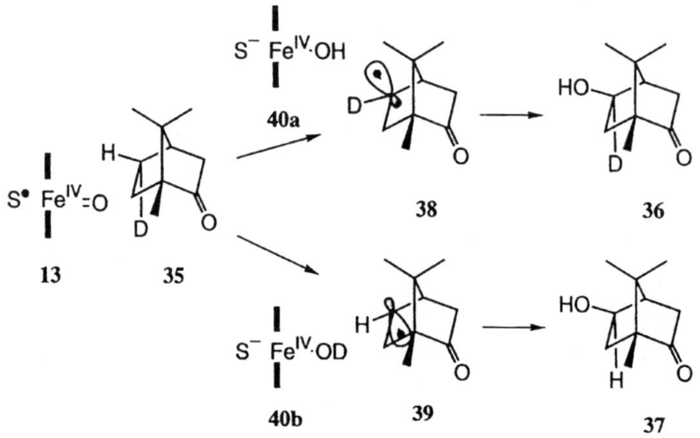

S⁻ Fe^IV·OH
40a
D
38
HO
D
O
36
H
S• Fe^IV=O
D
O
13
35
S⁻ Fe^IV·OD
40b
H
39
HO
H
O
37

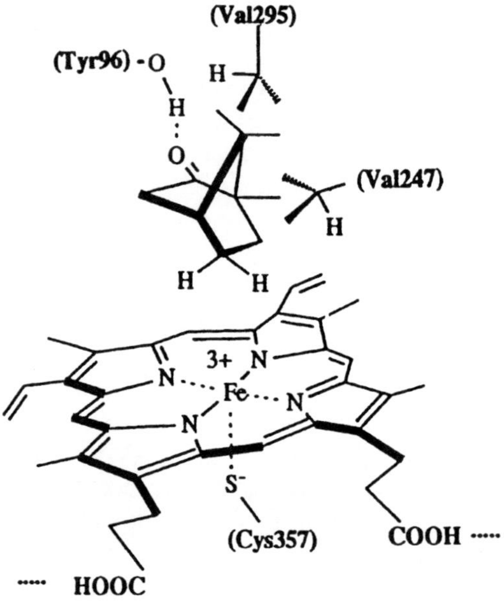

(Val295)
(Tyr96) - O
H
H
O
(Val247)
H
H
H
3+
N
N
Fe
N
N
S⁻
(Cys357)
COOH
HOOC

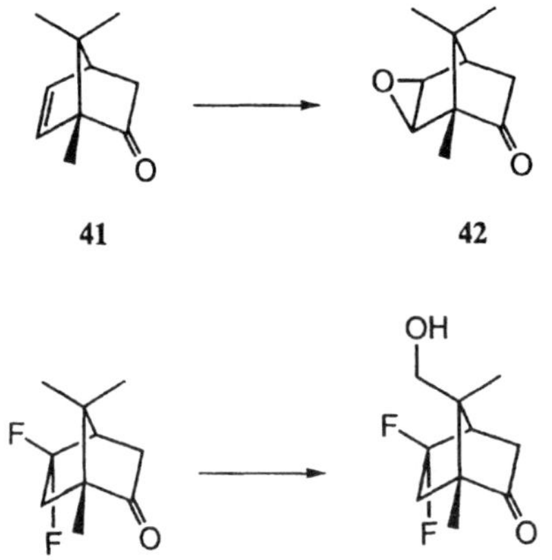

Fig. 9. Hydroxylation of substrate analogues by cytochrome P450$_{cam}$

situation suppressed intrinsic isotope effects are expected also due to nonlinear transition states for both *exo*- and *endo*-H˙ removal. Trapping of the carbon radicals **38** and **39** by HO˙, delivered from the hydroxy iron(IV)porphyrin **40**, then exclusively occurs from the *exo*-face of the molecule. An alternative mechanism can be discussed assuming homolytic O–O bond scission [47] of the protonated form of **9** (S⁻Fe(III)-O-O-H) to give HO˙, capable of removing both *exo*- and *endo*-H˙, and S⁻Fe(III)-O˙, which for steric reasons can trap the carbon radicals **38** and **39** only from the *exo* face. However, this sequence would exclude **13**↔**14** as reactive intermediates.

The preference for *exo* attack parallels the P450 catalyzed epoxidation of 5,6-dehydrocamphor **41** yielding **42** as the only product (Fig. 9) [48]. Another striking example for the tendency to expose the *exo* face is revealed by "metabolic switching" of hydroxylation from C(5) to C(9), when 5,5-difluorocamphor **43** is used as substrate [49]. The formation of **44** hydroxylated at C(9) can also be understood from the X-ray of P450$_{cam}$ because, except for C(5), the C(9) methyl group of camphor appears to be the C-atom closest to the oxygen binding site.

Concerning the metabolism of triterpenes and steroids, quite a number of P450 catalyzed transformations are very important, namely the 14α-demethylation of lanosterol [50], the side-chain cleavage of cholesterol [51] and pregnenes [52], and the desaturation of ring A of androgens with concomitant oxidative removal of C(19) [53]. The latter reaction is catalyzed by human placental aromatase, associated with a NADPH-dependent reductase, and requires three moles of oxygen and three moles of NADPH in order to oxidize androstenedione **45** to formic acid and estrone **46**, Fig. 10.

Fig. 8. Hydroxylation of camphor deuterated at C(5), and substrate binding in the active site of cytochrome P450$_{cam}$

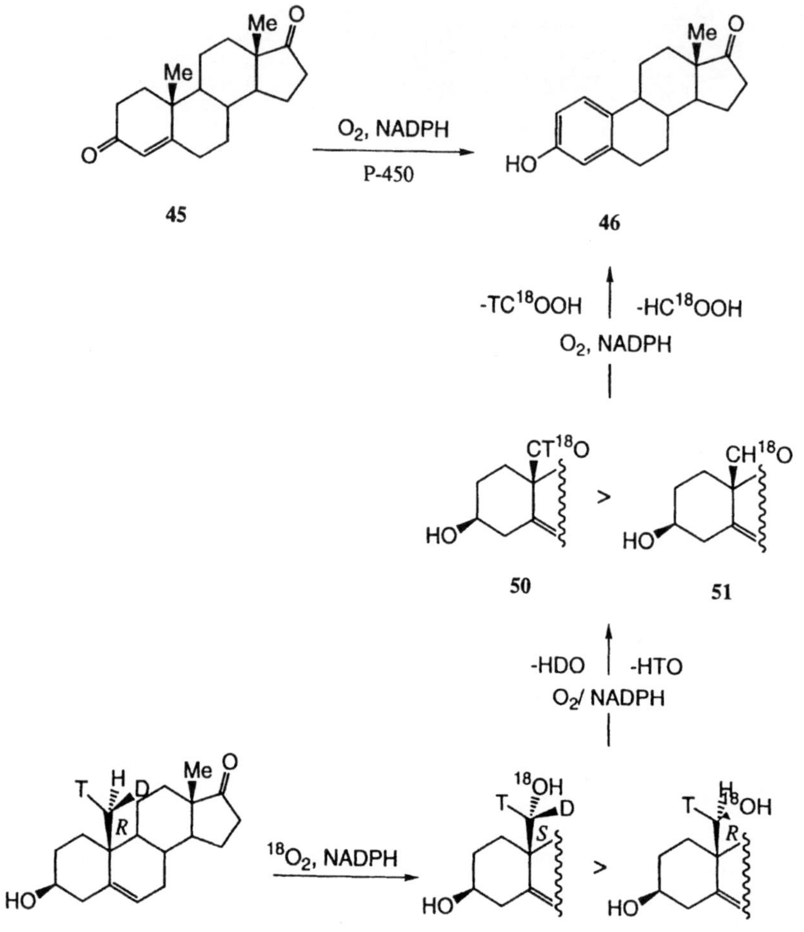

Fig. 10. Desaturation of ring A of androgens – the aromatase reaction

In view of the fact that the first step of the sequence displays a significant isotope effect, the stereospecificity of hydroxylation was investigated using the androgen (*19R*)-**47**, in which C(19) is chiral due to isotopic labelling [54]. Subsequent oxidation of the resulting alcohols (*19S*)-**48** and (*19R*)-**49** by exclusively removing the ^{2}H, and ^{3}H at C(19), respectively (corresponding to the *pro*-**R** hydrogen in the unlabelled substrate), yielded the aldehydes **50** and **51** and water. The third mole of oxygen is consumed in order to cleave the C(19)/C(10) bond and release the products **46** and formic acid. It is obvious that the stereospecificity of the hydroxylation at C(19) can be determined simply by measuring the ^{3}H-activity in water and formic acid: ^{3}H$_{formate}$ > ^{3}H$_{water}$ clearly indicated retention of configuration as shown in Fig. 10. Corresponding results were obtained incubating the (*19S*) diastereoisomer of **47**.

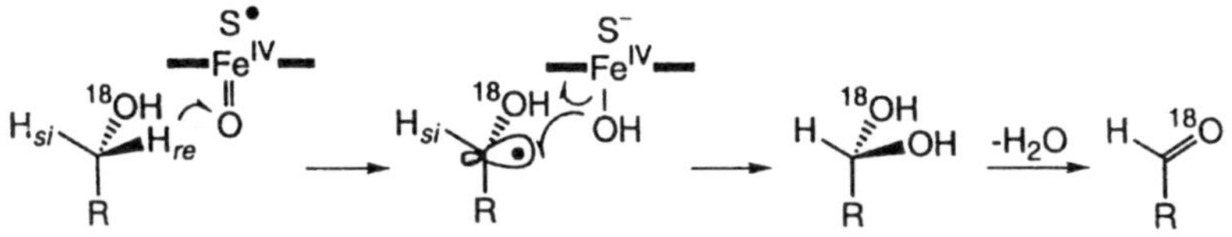

Fig. 11. Oxidation at the C(19) hydroxymethyl group

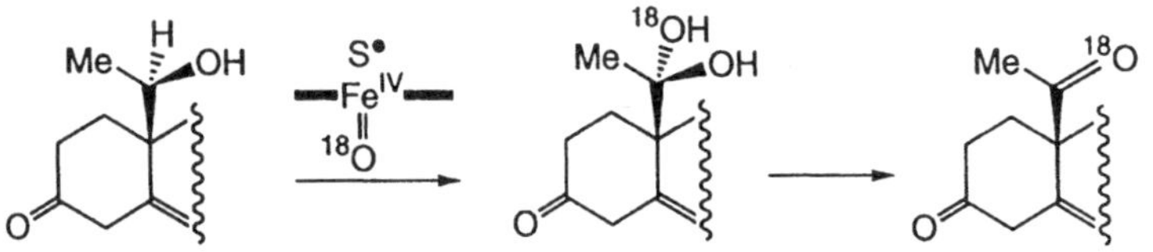

According to the reaction sequence in Fig. 10, the oxygen atoms of formic acid originate from the first and the third mole of O_2. Concerning the incorporation of ^{18}O, the most likely mechanism of a P450 catalyzed oxidation of an alcohol **52** to an aldehyde **53**, demanding removal of the *pro*-R hydrogen and retention of the ^{18}O-label, involves the stereospecific formation of the radical **54**, and radical recombination to give the *gem*-diol **55**, from which the (*pro*-R)-OH group is taken [55] (Fig. 11). This view is supported by the aromatase catalyzed oxidation of (*19R*)-**56**, a methyl homologue of **49** [56]. Incubation in the presence of $^{18}O_2$ resulted in up to 70% ^{18}O incorporation in the isolated ketone **57**. However, when the alcohol with (*19S*)-configuration was oxidized (^{18}O/aromatase) only 2.7% ^{18}O were found in the product **57**; a control experiment revealed that this figure could be attributed to exchange with the solvent. These results are consistent with the view that **55** and **58**, respectively, are valid intermediates in these P450 catalyzed oxidations (Fig. 11).

The remarkable C(19)–C(10) bond cleavage consuming the third mole of oxygen is still a matter of debate. It was shown that this reaction does not involve a 2β-hydroxy derivative [57] which was believed to de formylate C(10) via *hemi*-acetal formation [58]. The sequence shown in Fig. 12 is basically in agreement with published results and supported by the fact that the aldehyde **59** could be cleaved to the *tert*-alcohol **60** using a peroxoiron(III)porphyrin generated from *meso*-tetrakis(pentafluorophenyl)porphyrinatoiron(III)chloride and K_2O [59]. Treatment of **60** with aqueous 4N HCl furnished the estrogen **46**.

Accordingly, it can be assumed that the peroxoiron(III) intermediate **61** (see also **9**, Fig. 3) acts as a nucleophile on the aldehyde **59** to generate the adduct **62** which eliminates formate to give the hydroxyiron(IV) intermediate **40** and the

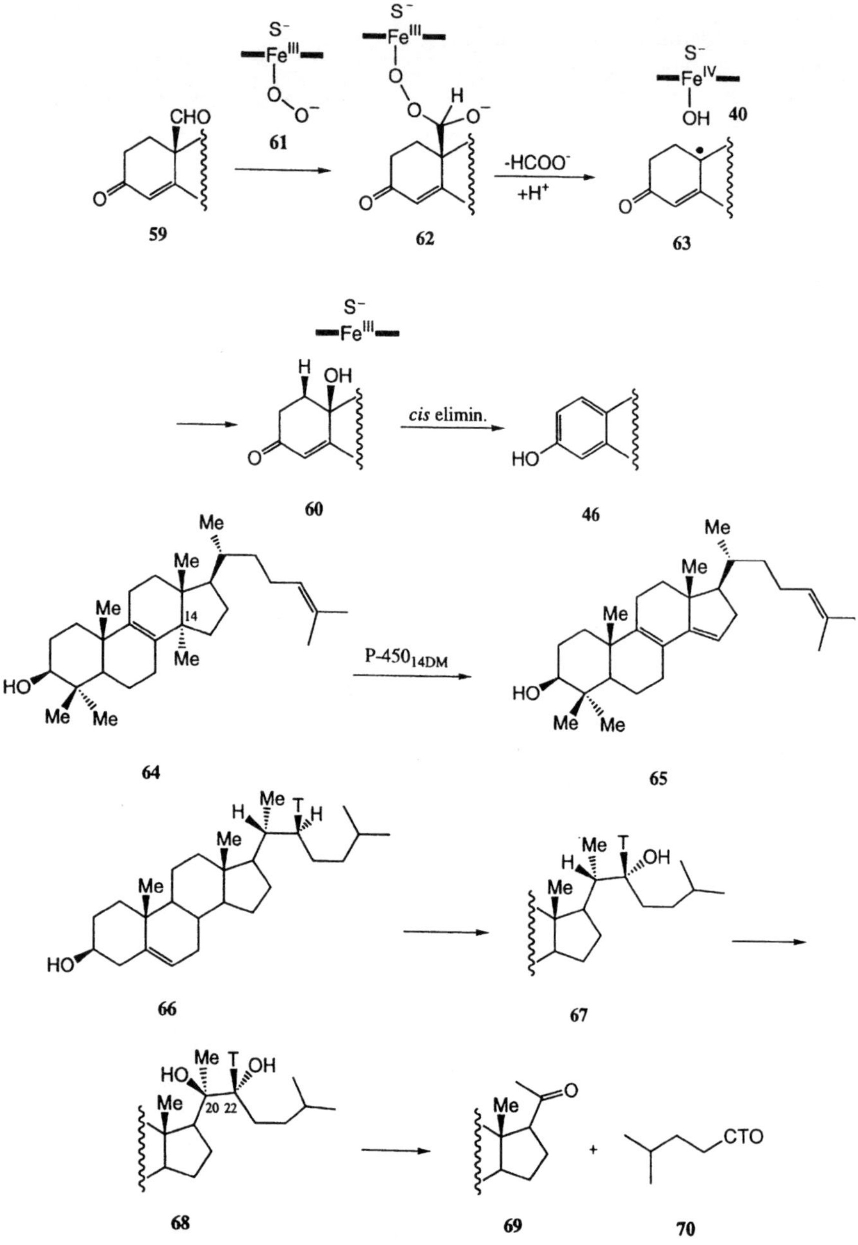

Fig. 12. Deformylation of C(19); 14α-demethylation of lanosterol **64**; side chain cleavage of cholesterol **66**

allylic radical **63**. Subsequent radical recombination yields the P450 resting state and the alcohol **60** which undergoes aromatization to **46** by *syn*-elimination of water (Fig. 12).

The participation of **61** and the formation of **62** in C–C bond scissons is also suggested by experiments with $P450_{LM2}/H_2O_2$ and cyclohexanecarboxaldehyde as the substrate [60]. Equimolar amounts of cyclohexene and formate were isolated only in the presence of H_2O_2, any other additive which usually gives the oxoiron(IV) intermediate (PhIO, mCPBA) failing to display the deformylation. Recent theoretical calculations [61] alternatively suggest that an unspecified nucleophile from the enzyme rather than **61** is adding to the aldehyde **59**. Subsequent enolization of the C(3) carbonyl followed by H⁺ removal from C(1) by the oxoiron(IV) intermediate could generate an allylic radical at C(1), facilitating C(10)–C(19) bond cleavage.

Except for aromatization, the 14α-demethylation of lanosterol **64** seems to operate by the same sequence of events in order to remove the angular C(14) methyl group with concomitant introduction of the C(14) double bond to furnish **65** (Fig. 12).

Concerning the side chain cleavage of cholesterol **66**, nothing conclusive is known about the mechanism of the actual C(20)–C(22) bond breaking step. The available information indicates that hydroxylation takes place first at C(22) under retention of configuration. Subsequent hydroxylation of the resulting alcohol **67** leads to the diol **68** which is cleaved to produce pregnelonone **69** and 4-methylpentanal **70** [51, 62].

1.3 Allylic Hydroxylations

From the incubation of the olefins **71–73** with rat liver microsomes, only mixtures of isomeric allyl alcohols could be isolated [63] indicating that cyclic allyl radicals are prone to rearrange significantly at the active site of cytochromes P450 (Fig. 13). A similar observation was made investigating the P450 catalyzed oxidation of (R)(+)-pulegone **74** to menthofuran **75** [64]. After incubation of **74**, trideuterated in the (Z)-methyl group, the isolated metabolite **75** was also found to be largely trideuterated, and became ^{18}O labelled in the furanoid oxygen if the experiment was carried out in an atmosphere of labelled oxygen (Fig. 13). A significant intramolecular isotope effect ($k_H/k_D = 7.7$) was determined from an incubation with the pulegone **76**, dideuterated in both allylic methyl groups. A mechanism consistent with these results involves the removal of H⁺ preferentially from the unlabelled (E) allylic methyl group of **74** to yield the allylic radicals **77**⇌**78**⇌**79**. Radical recombination of **79** with S⁻Fe(IV)-OH (**40**) gives the rearranged alcohol **80** which dehydrates to **75**.

In view of the already mentioned stereospecificity of O-insertion into nonactivated C–H bonds and the stability of allyl radicals, the observed "stereochemical scrambling" is somewhat surprising. In this context it is noteworthy to discuss a P450 reaction of the allylic substrate geraniol which is distinguished from those in Fig. 13 as it neither contains a ring system nor would intermediate radicals like **78** be involved [65]. Furthermore, reactions could be performed with an enzyme working on its native substrate. Geraniol is known to be an

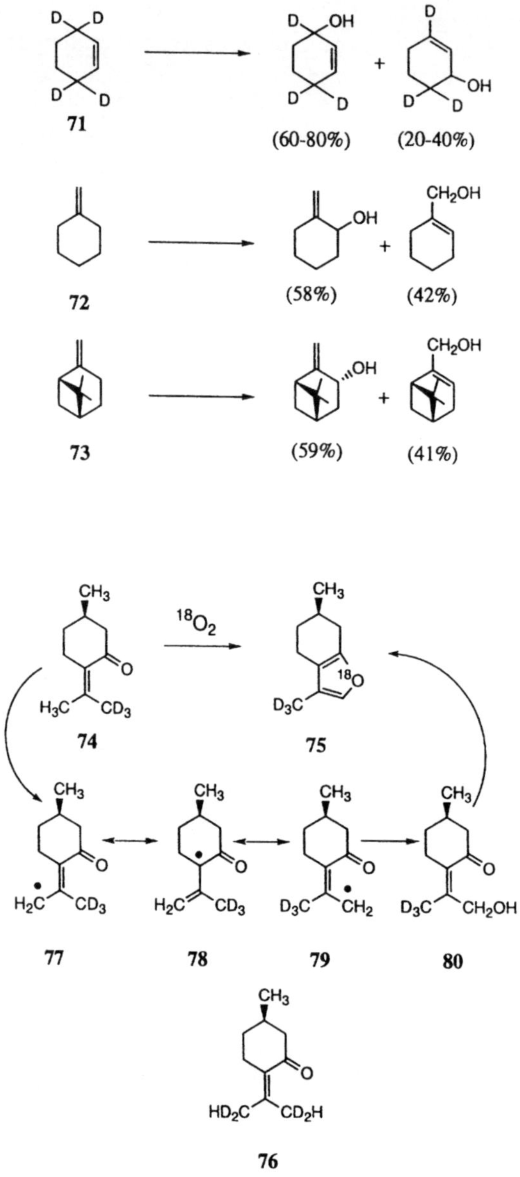

Fig. 13. P450 catalyzed allylic hydroxylations

early biosynthetic precursor of the terpenoid part of the plant indole alkaloids in which it is incorporated after multiple oxidations. Thus, when an enzyme fraction of P450$_{Cath.}$ was prepared from seedlings of the subtropical plant *Catharanthus roseus* and incubated with mg quantities of the deuterated and ^{13}C-labelled substrate **81** the 8-hydroxygeraniols **82** and **83** were the only products being recovered (40–45% yield) [65] (Fig. 14). Analysis of the [1]H- and

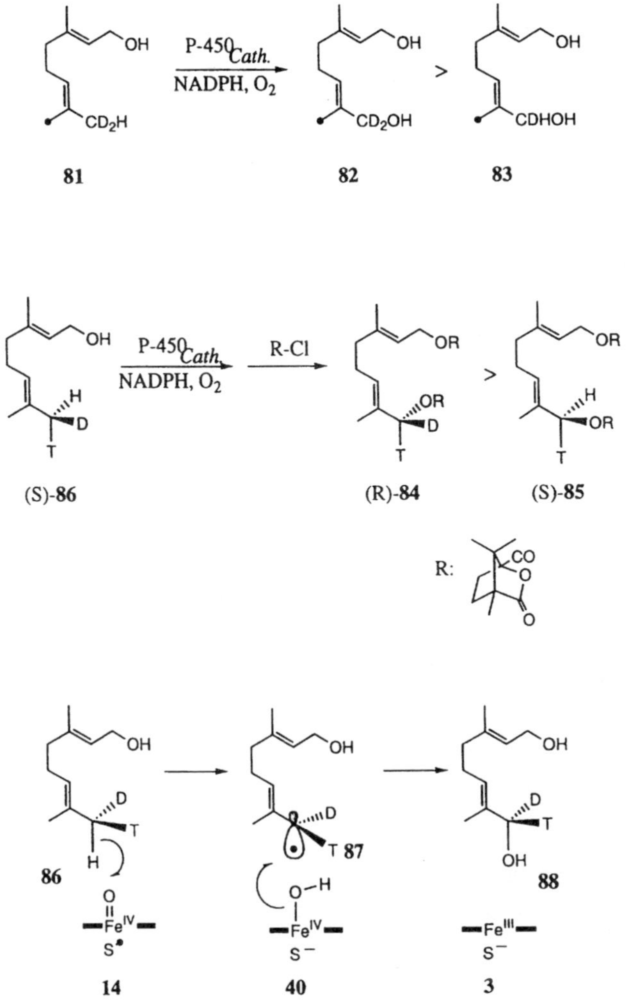

Fig. 14. Hydroxylation of geraniol of P450Cath.

^{13}C-NMR spectra revealed a substantial intramolecular isotope effect ($k_H/k_D = 8.0$) and retention of the ^{13}C-label at CH$_3$-C(7). This clearly indicates, in contrast to allylic systems so far investigated, a regiospecific H˙ removal from the (E)-methyl group of **81** followed by trapping of the allyl radical without rearrangement. Moreover, no isotope sensitive branching of the enzymatic reaction was observed when a geraniol sample fully deuterated in the terminal (E) methyl group was hydroxylated.

These results are a prerequisite in order to investigate the stereospecificity of the hydroxylation using geraniol samples which, due to isotopic labelling, are chiral at the allylic (E)-methyl group at C(7). The mixtures of chiral 8-hydroxygeraniols originating from incubations of geraniols of opposite chirality at

C(8) were analyzed by recording the ^{3}H-NMR spectra of the corresponding (−) camphanates [66]. These derivatives are suitable because the ^{3}H resonances of 8-(R)-**84** and 8-(S)-**85** are separated by 0.03 ppm. Correlation to reference samples of known chirality revealed that the down field signal corresponds to the (S)- and the high field resonance to the (R)-configuration at C(8). This method is equivalent, although less cumbersome, to the enzymatic analysis [67] by only considering the few molecules being chiral due to the ^{3}H-label.

From these experiments it became evident that the P450$_{Cath.}$ catalyzed hydroxylation of (S)-geraniol **86** gave a mixture of 8-hydroxygeraniols in which the (R)-dominates the (S)-enantiomer by 8:1 (Fig. 14). When (R)-geraniol was oxidized, (S)-8-hydroxygeraniol was the major product. Accordingly, the allylic hydroxylation of geraniol by P450$_{Cath.}$ proceeds with retention of configuration and with complete regio- and stereospecificity. This can be reasoned assuming that in the "cage" of the active site the allylic radical **87** is trapped by **40** much faster than any competing racemization process can proceed and hence **88** is obtained enantiomerically pure.

1.4. Epoxidations

The mechanism of epoxidation has been studied in detail both with P450 enzymes [68] and synthetic metal porphyrins [69]. The problem finding a conclusive answer on how the enzymatic reaction proceeds is due to the fact that intermediates have not been detected but inferred by investigating the stereochemistry of product formation. By and large it is safe to say that the reaction depends on the steric hindrance imposed by the olefin's substitutents, the electron donating character of the olefin, and the electron demand of the oxo-iron(IV) porphyrin used. In particular the last aspect makes it difficult to draw conclusions from reactions with model compounds, since these metal porphyrins behave quite differently from native P450 due to the distinct electronic nature of both the metal and the porphyrin.

Furthermore, the approach of olefins to the oxygen of the oxo-iron(IV) intermediate generated from *face protected* porphyrins is likely to be distinct from native P450 due to steric interactions.

The results obtained from experiments with P450 enzymes clearly indicate that (Z)-olefins, in particular electron rich ones, are better substrates than (E)-olefins and that epoxidation proceeds with retention of olefin stereochemistry. This implies a rate limiting approach of the olefin on a sideway trajectory with an angle probably $\geqslant 30°$ to the porphyrin plane in order to form a charge transfer complex with the electron deficient oxo-iron(IV) intermediate [70]. The latter has been described as having triplet oxenoid character due to mixing of oxygen p_x and p_y orbitals with d_{xz} and d_{yz} iron orbitals. A fast spin inversion of the $S^· Fe(IV)=O \rightleftarrows ^-S\text{-}Fe(IV)\text{-}O^·$ in the charge transfer complex would allow a concerted addition of the oxygen atom. This view is supported mainly by the calculated $\rho^+ = -0.93$ for the epoxidation of series of substituted

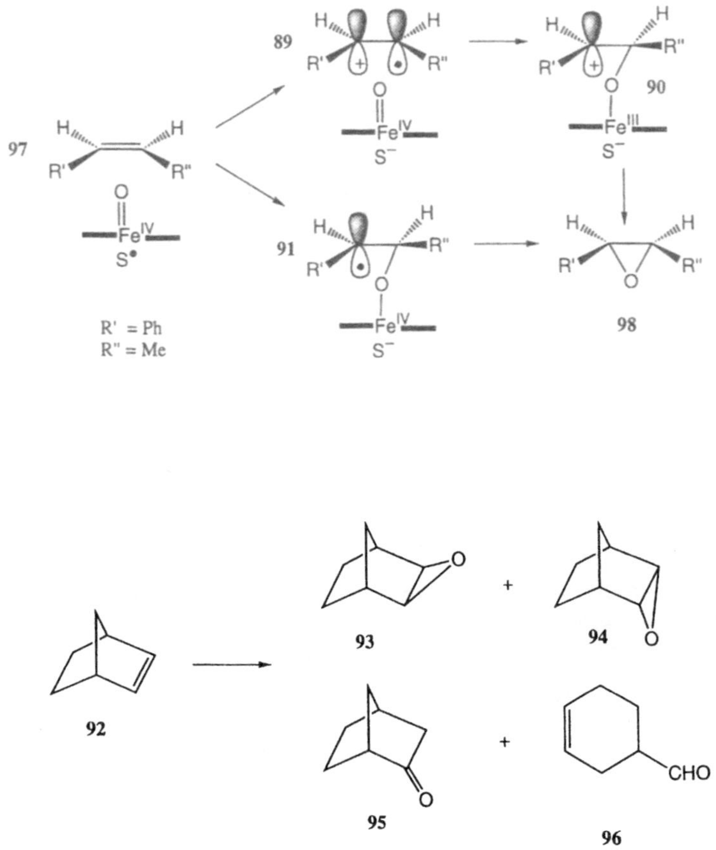

Fig. 15. Epoxidation of olefins

styrenes with ClFe(III)TPP/PhIO; in fact this value is quite similar to those reported for processes known to be concerted like carbene insertions into double bonds ($\rho^+ = -0.62$ to -1.61). However, experiments with P450 enzymes and with perhalogenated oxo-iron(IV)porphyrins suggest a stepwise reaction through intermediates **89–91**. For example the epoxidation of norbornene **92** with various model porphyrins and artificial "O" donors like iodosoxylene gave both *exo*- and *endo*-epoxinorbornane **93** and **94** in high yields [71]. The *exo/endo* ratio 3:1 parallels the results of the free radical chlorination of **92**. But since the epoxides were accompanied by up to 8% of the rearranged products **95** and **96** which are uncommon for a free radical pathway a carbocation radical like **89** was suggested to be the first intermediate in the epoxidation sequence (Fig. 15).

In contrast, the intermediate radical **91** was successfully used for AM1 calculations and molecular dynamics simulations (AMBER) in order to predict the P450$_{cam}$ catalyzed epoxidation of β-methylstyrene **97** [72]. Using the geometry of the known P450$_{cam}$ binding site, the re-face of the double bond was found to be the preferred orientation with respect to the ferryl oxygen. It also became clear that addition of the oxygen to C_β was favored by 9 kcal/mol over C_α due to the formation of the stabilized benzyl radical **91**. The results predicted 68% ee for the product, (1S, 2R)-**98** dominating, in good agreement with the experiment which gave 78% ee [73].

A stepwise reaction is also suggested by incubations of terminal olefins with P450 enzymes. It was discovered that these slow reacting substrates present a serious problem for P450 isozymes because frequently the heme became deactivated. The epoxidation of (E)-1-D$_1$-oct-1-ene **99** was studied in detail using P450 containing rat liver microsomes (Fig. 16) [68]. Incubation of **99** yielded exclusively the trans-epoxides **100** and **101** (5%), confirming a stereospecific addition of oxygen to the double bond. However, using undeuterated 1-octene the analysis of the corresponding 1-methoxy-2-octanols **102** and **103** revealed that the products were nearly racemic at C(2), (S)/(R) = 1.0:0.9. Accordingly, the enzymatic epoxidation proceeds with almost no discrimination between the re- and the si-face of the double bond (Fig. 16). Furthermore, in the course of the enzymatic reaction the P450 activity decreased by ~22%, and it was shown that the ring D nitrogen of the porphyrin became specifically alkylated by the substrate. This result was used in part to define the optimal substrate orientation above rings D, C, and A of the ironporphyrin [74]. Isolation of the N-alkylated heme **104** made it possible to determine the structure of the attached "epoxide" revealing exclusively the (S)-enantiomer **100**. Thus, whenever the olefin **99** exposes its re-face to the oxo-iron (IV) species the reaction is branched to give either N-alkylation or epoxidation Fig. 16. Such a sequence could involve a radical intermediate **105**, equivalent to **91**. Approaching the porphyrin with its si-face, no N-alkylation but epoxidation of **99** is observed. It seems that, in the latter case, C–N bond making is unfavourable due to steric or hydrophobic interactions of R with active site residues. In this context it is appropriate to discuss ironoxetanes like **106** and **107** as intermediates en route to epoxides; for steric reasons they are particularly attractive for terminal olefins. This view is supported by several kinetic studies using P450 mimics like oxomanganese porphyrins and hypochlorite [70, 75].

Concerning the substrate oct-1-ene **99**, the "normal" products **100** and **101** could be formed via [2$_a$ + 2$_s$] transition states, however, in the event of re-face attack, more severe interactions of R with the environment could open an alternative radical pathway leading to N-alkylation.

Ironoxetane intermediates are also supported by experiments with reconstituted P450$_{LM2}$ and (E)-1-D$_1$-propene **108** which unexpectedly afforded a mixture of 98% methyloxirane **109** and 2% trans-2-D$_1$-1-methyloxirane **110** [76]. This remarkable stereospecific H/D-exchange at the active site was confirmed in a control experiment by incubation of unlabelled propene in D$_2$O (80% D) which gave a mixture consisting of 80% **110** and 20% **109**. In agreement with these results is a mechanistic hypothesis (Fig. 17) involving the oxetane

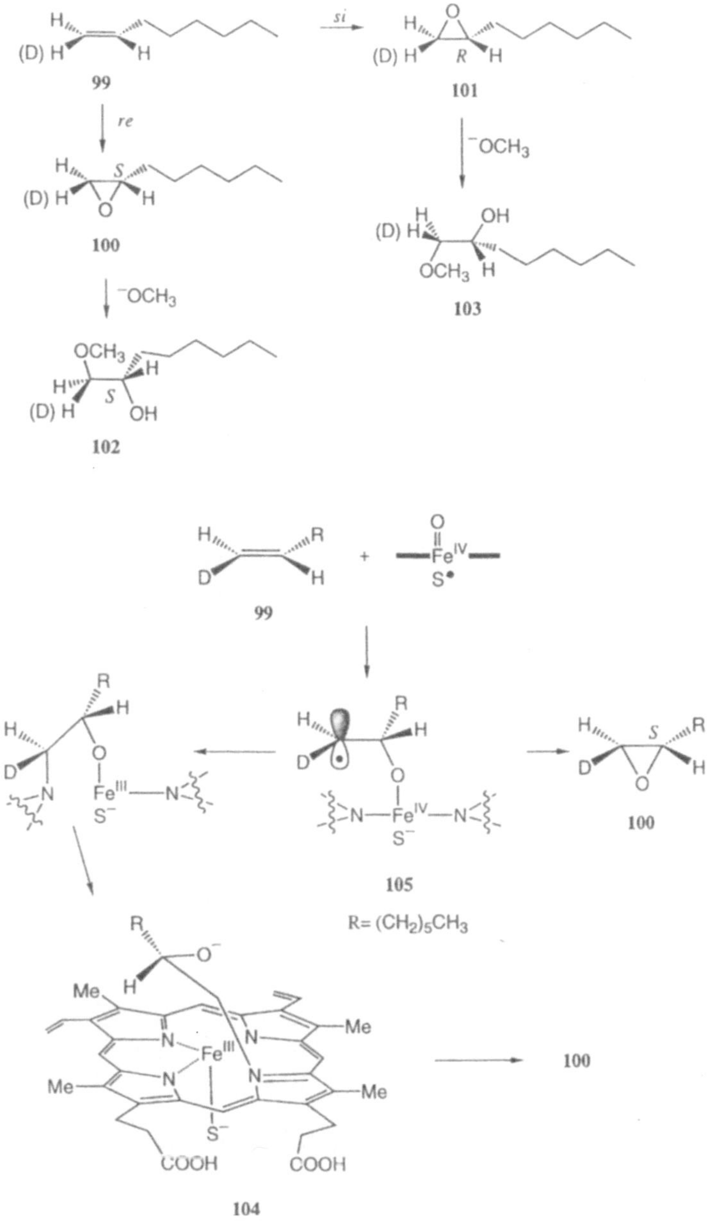

Fig. 16. Epoxidation of terminal olefins and suicide inhibition

intermediate **111**, from which a small amount of **110** is generated directly by reductive elimination. Stereospecific D-removal from **111** by an internal base affords the iron carbene **112**, subsequent ring closure to the epoxide **113** followed by protolysis of the Fe–C bond liberating the undeuterated product

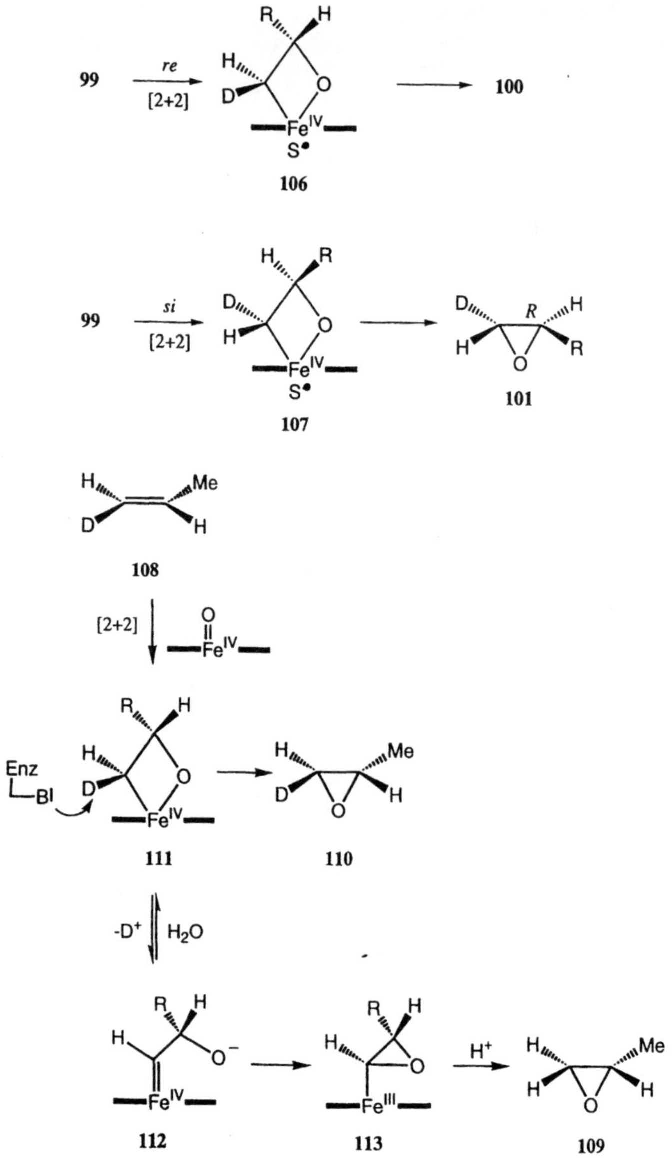

Fig. 17. Ironoxetanes in P450 catalyzed epoxidations

109. Deuterium incorporation from the solvent may occur in the last step and in the equilibrium **111** $\rightleftarrows$ **112**. The fact that epoxides originating from terminal $> C_4$ olefins show deuterium retention can be reasoned in two ways: either the formation of ironoxetanes is disfavored due to steric interactions or the internal base acting on the oxetane cannot operate due to steric constraints and consequently the long chain olefins have to depart from **111** by reductive elimination.

In conclusion it looks as if the reactive oxo-iron(IV) intermediate of the catalytic cycle of cytochrome P450 operates by different mechanisms depending on the structure and electronic nature of the unsaturated substrates.

1.5 Benzylic Hydroxylations

The mechanism of P450 catalyzed hydroxylations of benzylic carbons has been investigated both with enzymes and model compounds. Incubation of the deuterated, chiral ethylbenzenes 114 and 115 with $P450_{LM2}$ regiospecifically (99.8%) furnished a mixture of benzylalcohols 116–119 revealing that hydroxylation proceeds preferentially by H' removal in favor of the pro-S over pro-R hydrogen by about 4:1 (Fig. 18) [77]. Most strikingly the degree of inversion of configuration observed for pro-S hydrogen removal yielding the (R) benzylalcohols 117 and 118 is significantly higher ($\sim 40\%$) then for pro-R hydrogen abstraction ($\sim 20\%$) giving 116 and 119 displaying (S) configuration. Accordingly, in the course of enzymatic hydroxylation, H' removal and radical recombination are two clearly separated events with different selectivities.

Inversion of configuration is conceivable assuming that the benzyl radical 120 is not immediately trapped to give 116 but either undergoes rotation around the Ph-C' bond or is turning by 180° in the active site to yield the radical 121 which finally recombines to the product 117. Reorientation of the complete molecule is favored because π-stabilization by the aromatic ring is always preserved and, furthermore, the required mobility of the substrate in the active site is indicated by the fact that 1,1-D$_2$-ethylbenzene is hydroxylated in ortho and para positions of the phenyl ring to a greater extent than the undeuterated compound. This isotopically sensitive branching of the enzymatic reaction, in particular the formation of p-ethylphenol, suggests that $P450_{LM2}$ can tolerate quite different substrate orientations in the active site.

With the same chiral ethylbenzenes 114 and 115 catalytic hydroxylations were carried out using PhIO and the iron(III)tetraphenylporphyrin derivative 122 chiral with binaphthyl bridges spanning both faces of the porphyrin (Fig. 18) [78]. Interestingly, this catalyst has a twofold preference for removing the pro-R hydrogen of ethylbenzene almost stereospecifically with retention of configuration, e.g. reaction with 115 gave 117 in 87% yield. In contrast the removal of the pro-S hydrogen is accompanied with 20–25% inversion of configuration. The intramolecular isotope effects were found to be equivalent for both enantiotopic positions, $k_H/k_D = 6.4$ in contrast to the above-mentioned enzyme catalyzed reaction where a large isotope effect was estimated for the removal of the (favored) pro-S-hydrogen ($k_H/k_D > 10$) and a rather masked value was determined for pro-R hydrogen abstraction ($k_H/k_D = 4.0$) [77].

In order to determine intrinsic isotope effects of benzylic hydroxylations, the metabolism of different deuterated toluenes was investigated in detail with rat liver microsomes and compared with the chemical radical chlorination of toluene [79, 80]. Increasing deuteration of the methyl group of toluene led to a stepwise decrease in the formation of benzylalcohols in favor of cresols from 69% for unlabelled toluene to 24.3% for $CD_3C_6H_5$, clearly demonstrating isotopically sensitive branching of the enzymatic reaction. The ratio of cresol isomers, p-/o-/m-1:1:0.3 remained constant, but the total yield increased along

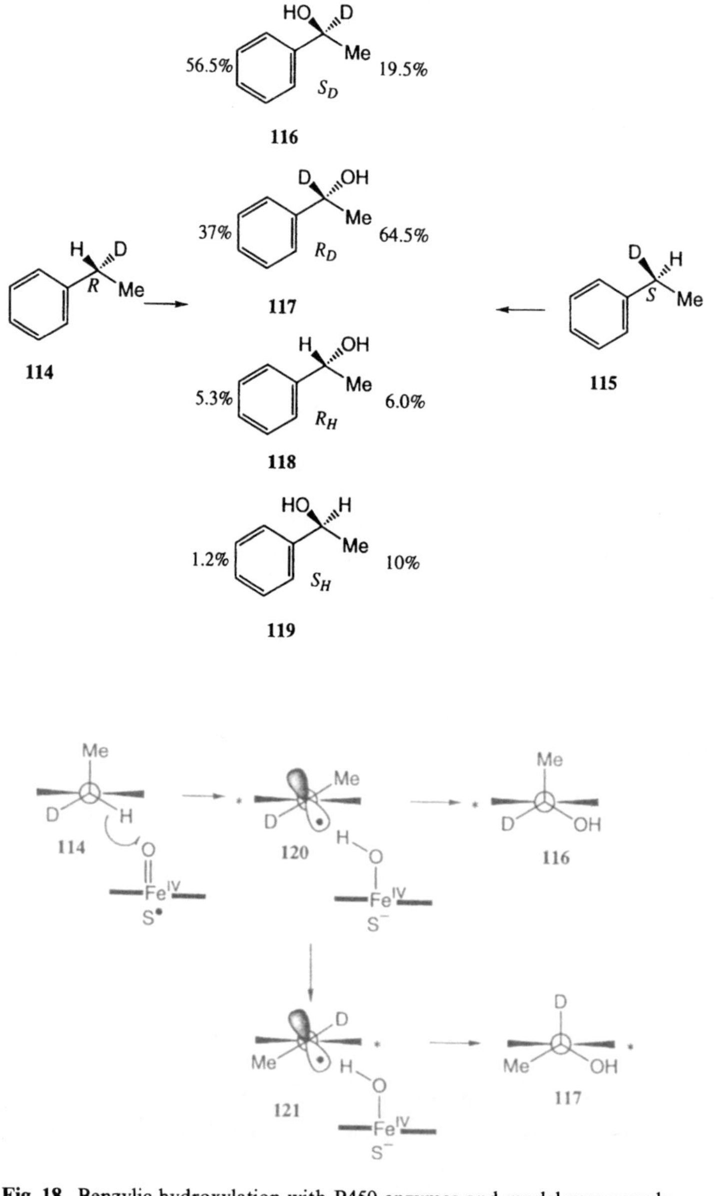

Fig. 18. Benzylic hydroxylation with P450 enzymes and model compounds

R:

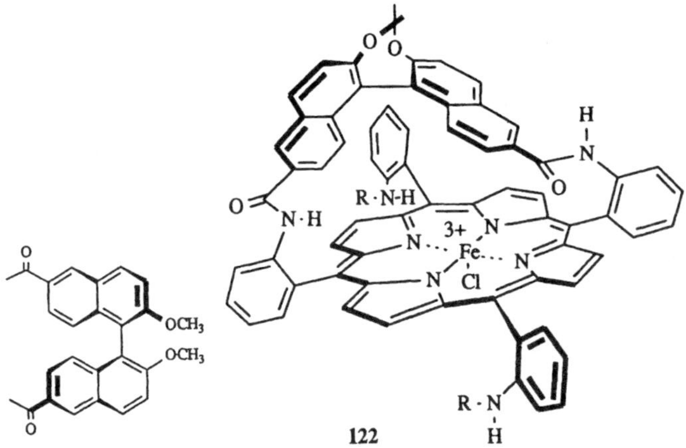

122

Fig. 18. Continued

stepwise deuteration of the methyl group. In contrast, deuteration of the benzene ring had no effect on product formation. The major difference between the P450 catalyzed hydroxylation and the chlorination of toluene is that for the latter all intramolecular isotope effects observed for toluenes with increasing deuteration are about the same $k_H/k_D = 5.90 \pm 0.41$, that is the removal of H$^\cdot$ or D$^\cdot$ from the methyl group is independent of the isotopic substitution at this carbon. Surprisingly, this is not the case for the enzymic reaction, and with increasing deuteration the intramolecular isotope effects (D_K) were found to increase from $k_H/k_D = 4.59 \Rightarrow 7.30$.

1.6 Aromatic Hydroxylations

The formation of carcinogenic products in the course of aromatic hydroxylation has initiated quite a number of mechanistic investigations [81]. The significance of arene oxides **123** which have been trapped and identified as intermediates in the formation of phenols was also supported by experiments with substrates like **124** deuterated in the position of hydroxylation. Incubation of **124** with hepatic P450 resulted in an unusual retention of deuterium in the product **125** and the migration of deuterium from the position of hydroxylation to the adjacent carbon, termed NIH shift [82], was taken as evidence that the arene oxide **123** could either be opened by protonation and rearranged to give **125** or pass through the ketone **126**, Fig. 19. The absence of a large primary isotope effect on the rate of hydroxylation determined for a number of substrates with rat liver microsomes is in agreement with both alternatives because the isotope sensitive step is not rate limiting. Thus, oxygen insertion into C–H bonds by H$^\cdot$ removal and subsequent radical recombination are rather unlikely. Consequently, two nonconcerted pathways were taken into consideration involving intermediates **127** and **128** reminiscent of those proposed for the mechanism of epoxidation

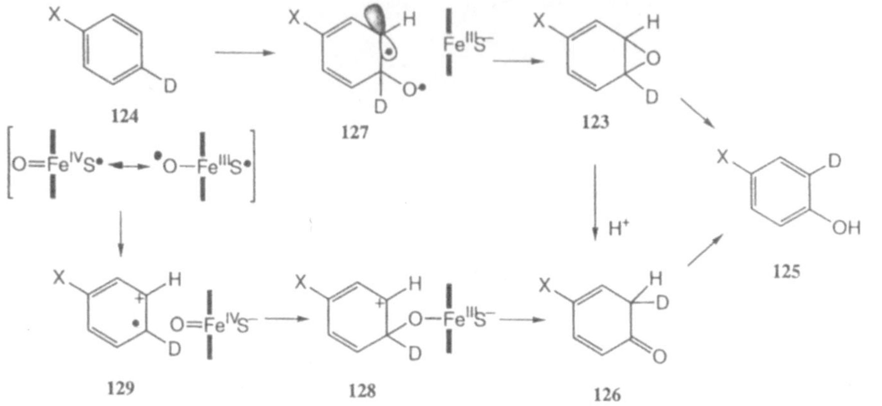

Fig. 19. Hydroxylation of aromatic compounds – the NIH-shift

(91 and 90, respectively, Fig. 15). According to theoretical calculations (MINDO) both sequences are distinct and mainly differ with respect to the development of positive charge [83]. In one case direct oxygen addition gives the biradical 127 leading to the arene oxide 123 whereas, alternatively, electron transfer from the benzene ring to the oxo-iron(IV) intermediate yields first the radical cation 129 and then 128 covalently linked to the iron porphyrin. The cation 128 either proceeds directly to the phenol 125 or via the ketone 126 rather than through the epoxide 123. The latter pathway is certainly more sensitive to environmental and substituent effects, which could stabilize the positive charge. It follows that electron donating groups in the substrate as well as charge stabilizing residues near the active site would favor a route to phenols circumventing arene oxide formation. The significance of arene oxides as obligatory intermediates is also questioned by results obtained from microsomal hydroxylation of a series of deuterated monosubstituted benzenes [84]. Examining the regioselectivity of hydroxylation, the deuterium content in the products and the isotope effects it was claimed that hydroxylation proceeded via 126 rather than 123 indicated by an uniform $k_H/k_D = 4.05$. Moreover, the results could only be explained [85] assuming that, at least in part, insertion of oxygen was occurring "directly" by complete loss of deuterium from the site of hydroxylation and not involving arene oxides 123 or cyclohexadienones 126 as intermediates. In conclusion, the available data support a branched pathway, one most likely passing through intermediate 126 and, in parallel, another "direct" route departing e.g. from 128.

1.7 Oxidations Involving Heteroatom Participation

Cytochrome P450 enzymes not only oxidize nitrogen- or sulfur-containing substrates at the heteroatom to afford the corresponding N-oxides and (chiral)

sulfoxides [86] but are also capable of oxidative dealkylations. In particular oxygen-containing substrates undergo the latter reaction exclusively. The O-deethylation of 7-ethoxycoumarin **130** has been studied in detail [87, 88] revealing intrinsic isotope effects $D_k = 13-14$, depending on the P450 source. Furthermore, significant metabolic switching was observed, since besides **131** the product of unlabelled substrate 6-hydroxy-7-ethoxycoumarin **132** was isolated, and the product ratio **131**:**132** changed from 20:1 to 2:1 with increasing deuteration. Incubating 7-ethoxycoumarin chiral due to a stereospecific tritium label at the methylene group revealed that the *pro*-S-hydrogen is removed with > 95% selectivity [88]. Accordingly, cytochrome P450 attacks O-alkyl groups analogous to nonactivated C–H bonds by rate determining, stereospecific H˙ abstraction. The P450 catalyzed oxidative cleavage of esters [89] displays the same characteristic; hydroxylation proceeds at the carbon adjacent to oxygen, and intrinsic isotope effects as high as $D_k = 15$ were determined. In contrast, intrinsic isotope effects associated with the N-dealkylation of amines were generally found to be much smaller, $D_k < 3$ [90]. Interestingly, the magnitude of these isotope effects corresponds closely to those measured on similar substrates with chemical ($KMNO_4$, ClO_2), photochemical, and electrochemical methods [91]. Since for the latter reactions it is known that one electron oxidation at nitrogen is the first step, it has been suggested that electron removal from nitrogen is the first and rate limiting step in the P450 catalyzed demethylation of amines. For example, this would explain the small intrinsic isotope effect $D_k = 1.7$ found for the oxidation of N,N-dimethylaniline **133** deuterated in one methyl group with two different P450 isozymes [90]. After the slow formation of the nitrogen radical cation **134** one of the methyl groups is deprotonated to yield a carbon radical, e.g. **135**, which recombines with the hydroxylated iron(IV)porphyrin to furnish the carbinolamine **136** hydrolyzing to the secondary amine **137** and formaldehyde. As shown in Fig. 20, the latter becomes ^{18}O-labelled, if the experiment is carried out $^{18}O_2$ [92, 93]. Thus, comparing the results of enzymatic N- and O-dealkylation it can be concluded that differences in electronegativity are reflected in the preference for abstracting either an electron from the heteroatom followed by deprotonation or removing H˙ from the adjacent carbon. The analogy between electrochemistry and P450 catalysis becomes less clear when N,N-dimethylamides are oxidized. For the enzymatic reaction [94] intrinsic isotope effects as high as $D_k = 6.9$ were determined whereas the anodic oxidation revealed distinctively smaller values (2.16–2.78) [95]. These results suggest that the enzymatic process is likely to proceed in favor of H˙ removal from the carbon adjacent to nitrogen. The electrochemical reaction, however, involves electron removal from the nitrogen as the first and rate limiting step.

Further support for electron abstraction from nitrogen is derived from experiments on 1,4 dihydropyridine **138** with hepatic microsomes [96]. In the course of the incubation significant deactivation of the P450 enzyme was observed suggesting heme alkylation. Subsequent isolation and characterization of N-ethylprotoporphyrin IX indicated an ethyl radical transfer from the

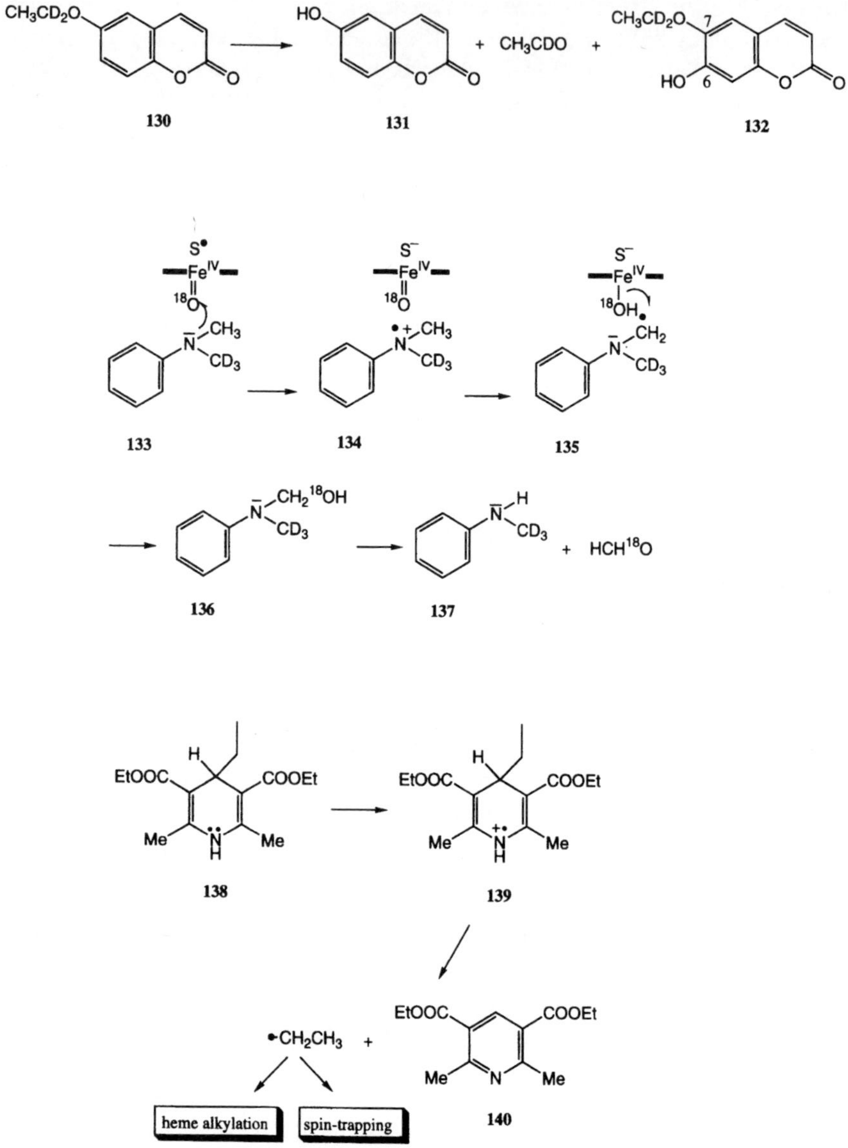

Fig. 20. Oxidative dealkylation at oxygen and nitrogen

substrate to the porphyrin; this was confirmed by spin-trapping the radical with α-(4-pyridil-1-oxide) N-*tert*-butylnitrone. The formation of an ethyl radical is evidence that a one-electron abstraction from the nitrogen of **138** generates a resonance stabilized radical cation **139**, which on releasing the ethyl radical from C(4) can aromatize to **140**.

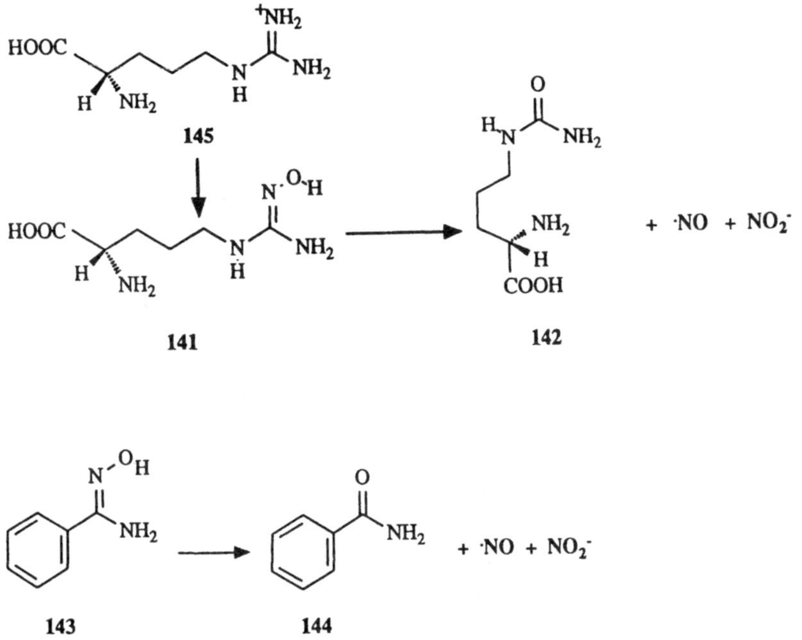

Fig. 21. NO-synthase reaction

An interesting P450 reaction has been reported recently [97]. It has been found that liver microsomes, in particular those from dexamethason pretreated rats (3A subfamily induced), were able to perform a reaction characteristic of NO-synthase [98], also a heme thiolate protein, and convert N^ω-hydroxy-L-arginine 141 into citrulline 142, Fig. 21. Evidence for the concomitant formation of ·NO rests on the spectroscopic identification of the corresponding NO-heme complexes and NO_2^-. The enzymic reaction requires O_2/NADPH, is inhibited by CO, and shows a significant rate decrease due to the inactivation of P450 by ·NO. Similar enzymatic activity and reaction characteristics were observed when the arylamidoxime 143 was used as a substrate [99] leading to the formation of the amide 144 and ·NO. The release of ·NO from substructures like amidines and amidoximes raises the question concerning the toxicological and pharmacological effects of drugs containing these functions. An equally important, although, unsolved, problem concerns the mechanism of this C–N bond cleaving reaction. In the absence of direct evidence it seems likely [100] that S^--Fe(III)-O-O$^-$ 61, Fig. 12, acts as a nucleophile on 141 (from which H· has been removed) resembling the mechanism of the C–C bond breaking step in the aromatase reaction. 141 is an established intermediate in the NO-synthase catalyzed sequence 145 $\Rightarrow$ 142 and produced by common P450 N-oxidation of L-arginine 145.

2 Active Site Analogues of Cytochrome P450

2.1 Introduction

Studies with synthetic P450 enzyme models have already been mentioned to contribute to the understanding of the mechanisms of P450 action and to provide information on the possible structure and electronic nature of iron porphyrins as intermediates of the catalytic cycle.

The second part of this review is concerned with synthetic aspects of those iron porphyrins which not only present spectroscopic equivalents of the catalytic cycle but, at least potentially, are capable of simulating reactions characteristic of cytochrome P450. In principle this can be accomplished with rather easily accessible iron-*meso*-tetraphenylporphyrin derivatives with "O" donors such as PhIO, NaIO$_4$, H$_2$O$_2$ or ROOH, [101], see "shunt-pathway", Fig. 3. However, problems with the stability of the porphyrin chromophore in the presence of powerful oxidants and the difficulty in controlling the oxidation- and spin states of iron have induced many research groups to employ *face protected*, and/or *electron deficient* iron porphyrins, even ligands complexing metals other than iron. A further challenge is the regio- and stereospecificity of P450 oxidation which requires a strict orientation of the substrate. In the most advanced model compounds the protein as a modulator of P450 reactivity and important to substrate binding has been replaced by artificial recognition sites which bind the substrates noncovalently.

Before discussing these developments in detail one has to acknowledge the outstanding efforts of quite a number of research groups who have investigated oxygen binding to synthetic iron porphyrins and hence provided the chemical background for the synthesis and application of P450 models. Very early it was recognized that iron(II)porphyrins unprotected on at least one of their faces reversibly bind oxygen to yield an oxygen bridged dimer which, after homolytic cleavage of the O–O bond, reacts irreversibly with the starting material to give a μ-oxodimer (Fig. 22).

In order to avoid such reactions several concepts for preparing porphyrins sterically hindered on one or both faces have been developed, nicknamed "capped" [102], "strapped" [103], "picket fence" [104], "bridged" and "doubly bridged" [105] or "basket handle" [106] porphyrins. Due to its relatively easy synthetic access, the "picket fence" approach in various modifications has become very popular in preparing P450 mimics. The general idea of this concept is that, for *meso*-tetraphenylporphyrin (TPP) derivatives like **146** with pivalyl groups in *ortho* positions of the phenyl rings, the activation energy for rotation around the *meso*-C/phenyl-C bond is large enough to restrict the interconversion of atropisomers at 25° in solution. Accordingly, if one face of the porphyrin is shielded by a base (B) the opposite site provides a "picket fence" pocket for oxygen binding. A simple modification of the synthesis leads to metal complexes where different ligands can be attached to the metal. Thus, selective acylation of

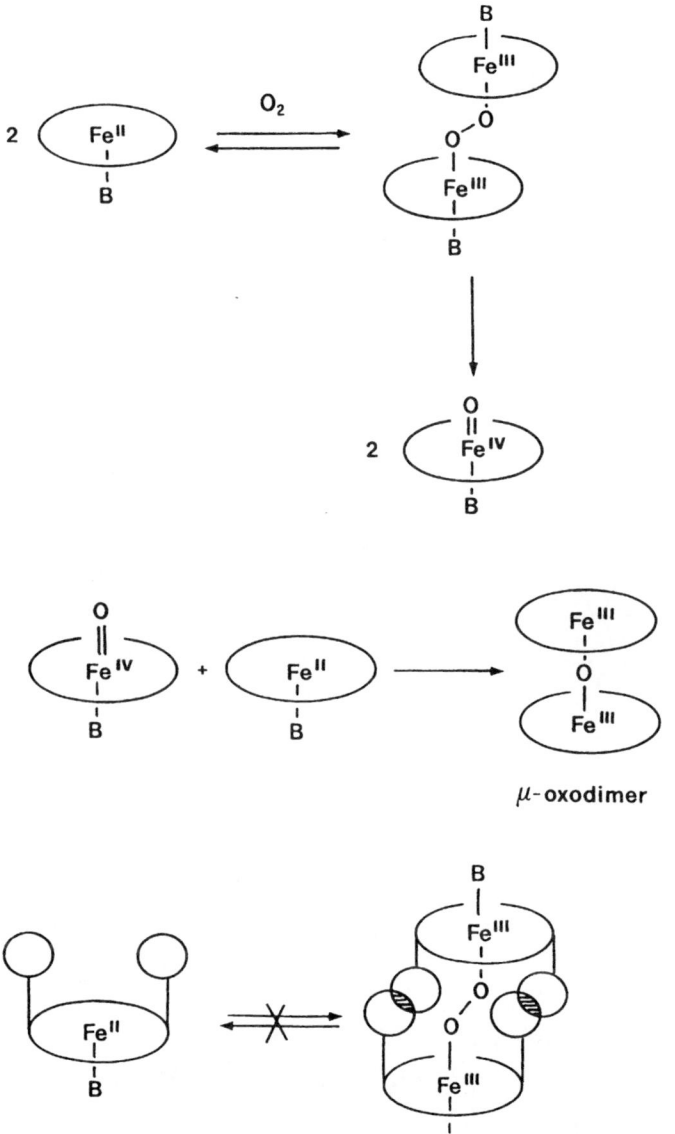

Fig. 22. Formation of μ-oxodimers by iron(II)porphyrins in the presence of O_2

the *meso*-tetrakis(*o*-aminoporphyrin) **147** yields **148**, carrying one unprotected aminophenyl group which can be thermally isomerized to furnish the α,α,α,β-porphyrin **149**. Subsequent modification of the "β-handle" conveniently provides the fifth ligand for iron, see **150**, Fig. 23 [104].

A more rigid orientation of additional ligands to the metal can be accomplished with the "*doubly bridged*" or "*basket handle*" approach. Furthermore, by

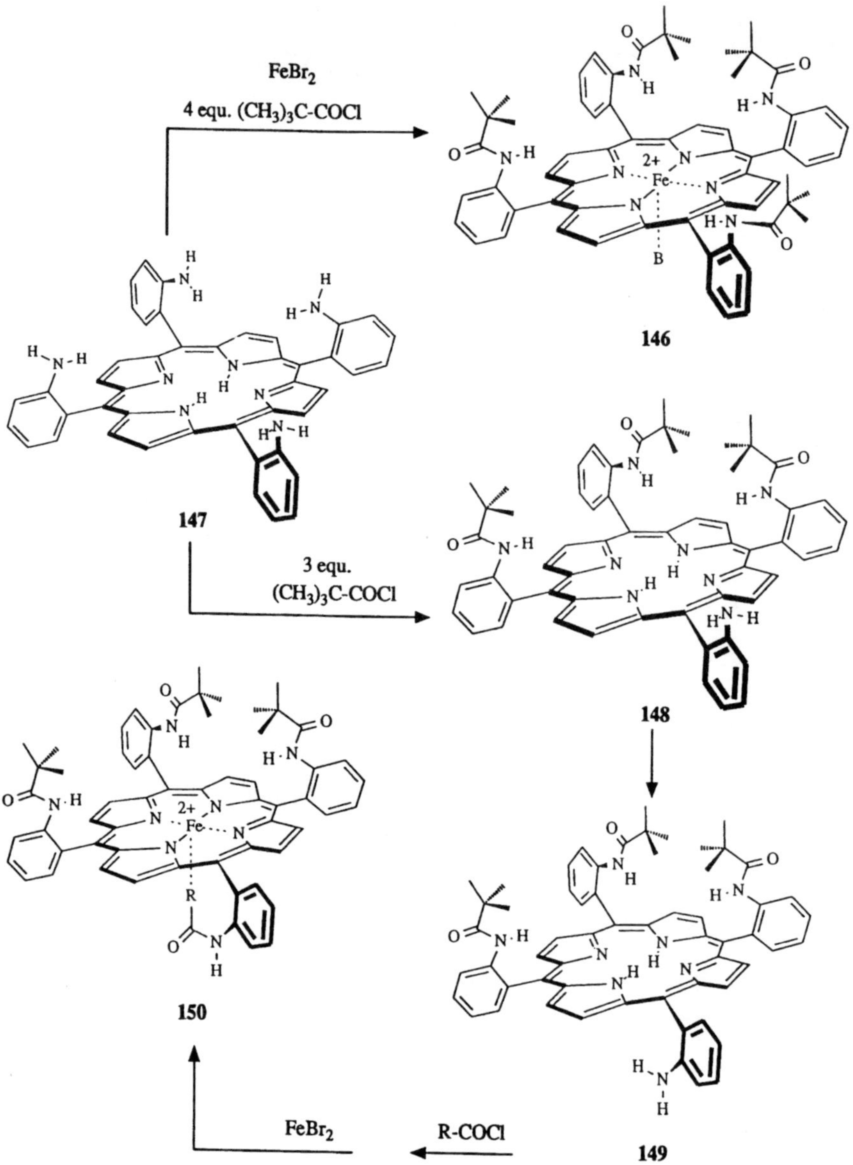

Fig. 23. Synthesis of *picket fence*-porphyrins

choosing suitable bridges, the nearest environment of the porphyrin can be modified concerning the lipophilic and hydrophilic character, respectively. The synthesis of metal porphyrins with identical bridges **151** is accomplished by condensation of an α,ω-dibromoalkane **152** with a phenolic porphyrin **153** under high dilution in the presence of Cs_2CO_3 (Fig. 24) [105]. If a porphyrin

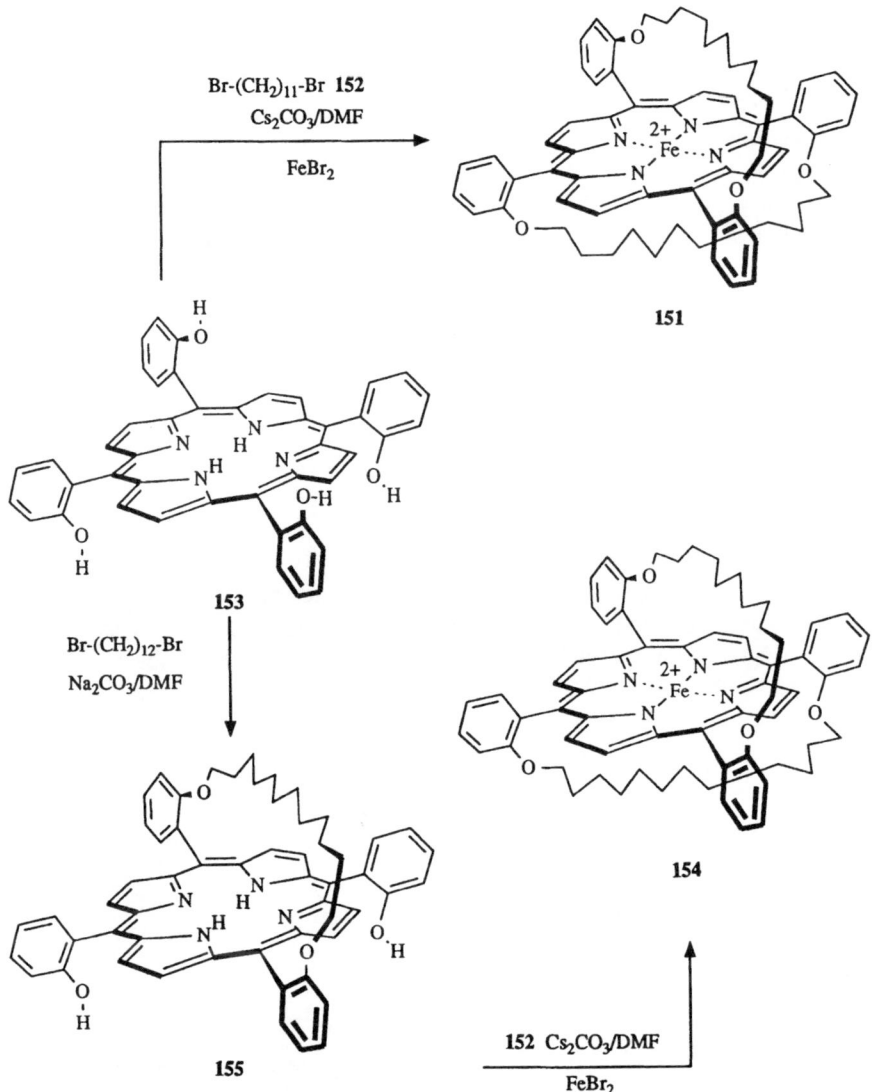

Fig. 24. Synthesis of mono- and dibridged porphyrins

with different bridges **154** is required, the preparation of the monobridged chromophore **155** can be achieved in up to 60% yield using either K_2CO_3 [106] or preferable Na_2CO_3 [105]; the second bridge is then attached under Cs_2CO_3 catalysis [105].

Except for a few examples [103], see below, the focus of synthetic P450 chemistry has been on the *meso*-tetraphenylporphyrin system mainly because free *meso*-positions can easily be oxidized in the presence of "O" donors employed in the "shunt pathway".

2.2 Thiolate Coordinating Iron Porphyrins

Synthetic P450 analogues most closely related to the active site are certainly those containing a thiolate coordinating to the iron. However, rather few attempts have been made to prepare such iron porphyrins due to the difficulties imposed by the presence of thiol/thiolate as the fifth ligand.

Recently, syntheses of the model compounds **156** and **157** were reported [107, 108], which are closely related to earlier approaches [103, 109] (Fig. 25). In agreement with theoretical calculations the CO complexes of the Fe(II)porphyrins **156** and **157** display a split Soret band at 370/446 nm and 383/456 nm, respectively, but no experiments with molecular oxygen were reported. But it was demonstrated that **157** catalyzed the formation of stable aryloxy radicals from the corresponding phenols in the presence of e.g. *tert*-butylhydroperoxide (TBHP). These results indicate a thiolate mediated O–O bond cleavage of TBHP accelerated ∼240 fold in comparison to iron(III)tetraphenylporphyrin [108].

In order to obtain a thiolate carrying P450 mimic capable of cleaving molecular oxygen it seems appropriate to "*force*" the thiolate into coordination with the metal through steric congestion. This idea led to the design of the cytochrome P450 model **158**, which has the thiolate ligand connected to a rather short bridge spanning the porphyrin face [105]. Motion of the thiolate is restricted by the ortho connection at the aromatic ring and the presence of the *tert*-butyl group. **158** is an equivalent of the E·S complex of P450 since the

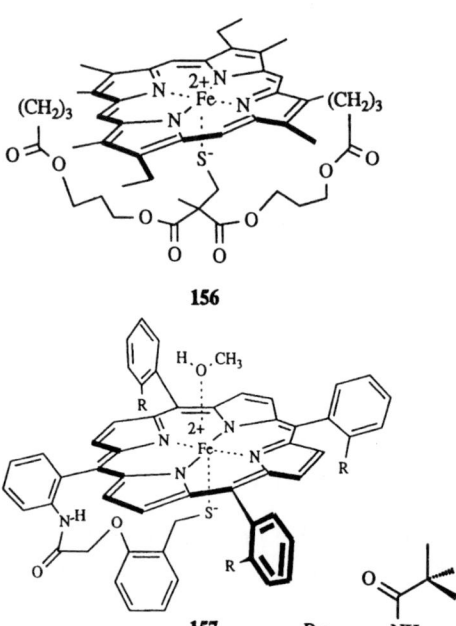

156

157   R :   NH

Fig. 25. P450 models with a thiolate ligand attached to the porphyrin

aliphatic chain bridging the porphyrin on the face opposite to the thiolate ligand is supposed to be hydroxylated under conditions suitable for generating the iron-oxo intermediate.

For the synthesis of **158** the monobridged porphyrin **159** was coupled with the dibromide **160** containing the sulfur protected as a thiocarbamate (Fig. 26). The resulting free base **161** was treated with $FeBr_2/THF$ to yield the high spin Fe(III) complex **158**. On the other hand, deprotonation of **161** with KH followed by $FeBr_2$ directly generated the iron(II)porphyrin **162** (Soret band 422 nm) which, on the addition of CO, gave two bands of equal intensity at 457 nm and 406 nm in the UV-/vis spectrum. The corresponding ^{13}CO complex **163** revealed a ^{13}C resonance at 206.6 ppm, close to the ^{13}CO complex of cytochrome P-450$_{cam}$ (203.3 ppm) [110]. The ^1H-NMR spectrum of **164**, (HS-Fe(II)) the protonated form of **162**, is characteristic of an intermediate Fe(II) spin system (S = 1) and shows remarkable changes of isotropic shifts in comparison to iron(II)porphyrins lacking the thiol ligand. Moreover, the extremely high-field shift of the SH proton (-75.9 ppm) indicates its close contact to unpaired spin density of the iron(II)-d orbitals [105]. Cyclic voltammetry in DMF/0.1 M $LiClO_4$ revealed that the redox potential of Fe(III) $\rightleftarrows$ Fe(II) is -607 mV (vs NaCl-SCE; $\Delta E_{anod/cath} = 78$ mV), [11, 110]. The addition of one equivalent of pentafluoro iodosobenzene to the E·S mimic **158** furnished the monohydroxylated high spin iron(III)porphyrin **165** in 77% yield (Fig. 26) [28]. The same product was isolated using m-chloroperbenzoic acid or tert-butylhydroperoxide. Treatment of **158** with a small excess of $^{18}O_2$ gave ^{18}O-labelled **165** accompanied by a small amount of sulfonate. The oxidation at sulfur could be prevented working in the absence of light [110].

Thus, it was demonstrated for the first time that a thiolate carrying P450 model is capable of cleaving molecular oxygen and inserting one oxygen atom into a non activated C–H bond. This is clear evidence for the intrinsic reactivity of such a system, although, since no reducing agent was added, the reaction is more reminiscent of suicide inactivation of native P450 in the absence or in case of slow electron delivery than proceeding through the "normal" catalytic cycle. A mechanistic proposal involving H· removal from the solvent toluene and homolytic cleavage of the O–O bond has been published [28, 110]. In order to perform catalytic, P450-like reactions the enzyme model **158** has recently been modified, see **166**, Fig. 27, providing "*Kemp acids*" as substrate recognition sites, [111].

Recent mutagenesis techniques have been used to generate mutants of human myoglobin *inter alia* such that has the axial histidine ligand (His93) replaced by cysteine (MbHis93Cys) [112]. The resulting iron(III)protein MbHis93Cys displayed spectroscopical features (UV, ESR, resonance Raman) almost identical to the high spin form of P450 enzymes. The redox potential ($E^0 = -230$ mV) is also close to the natural iron proteins. However, resonance Raman spectroscopy of the corresponding high spin iron(II)-Mb-His93Cys revealed that a distal His rather than the Cys thiolate is coordinating to the iron. This is also evident from unsuccessful attempts to prepare the CO complex with the characteristic split Soret band; in fact the OC-Fe(II)-MbHis93Cys

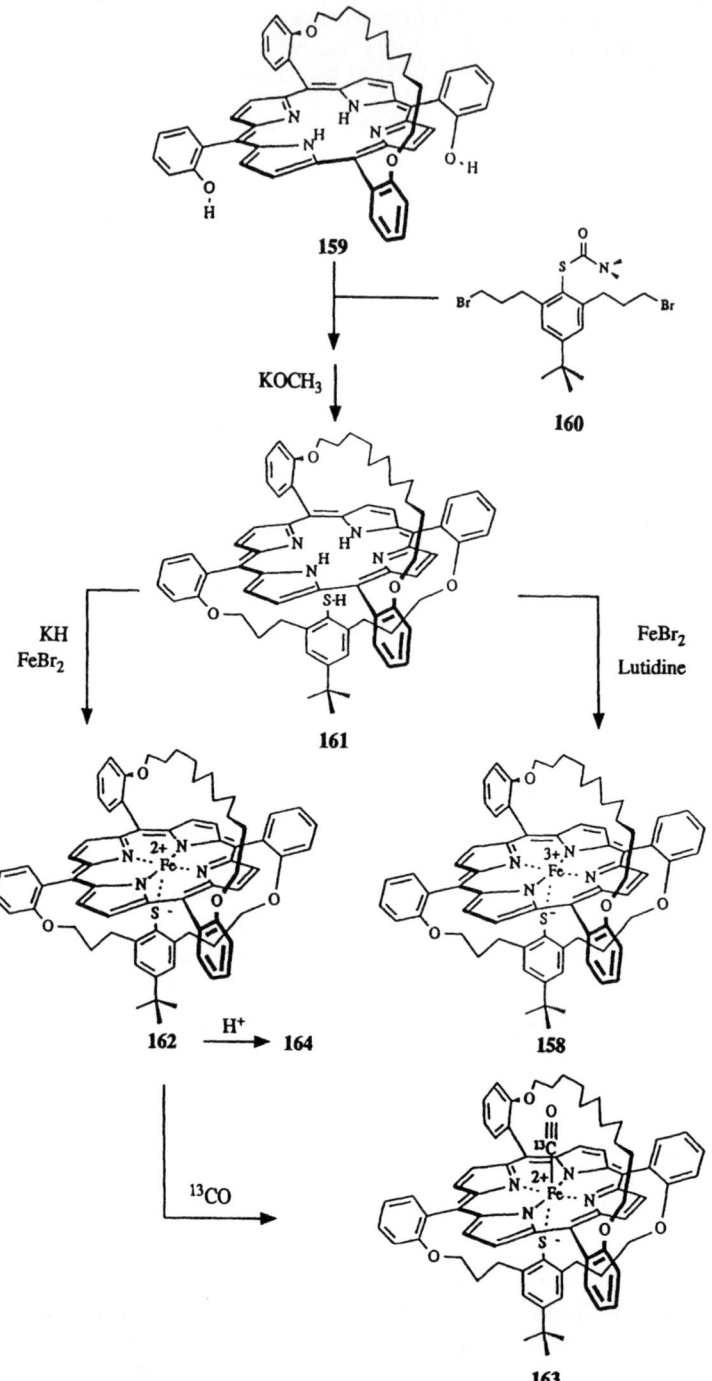

Fig. 26. Synthesis of a doubly bridged ironporphyrin carrying a thiophenolate ligand – a P450 model with respect to spectroscopy and chemical reactivity

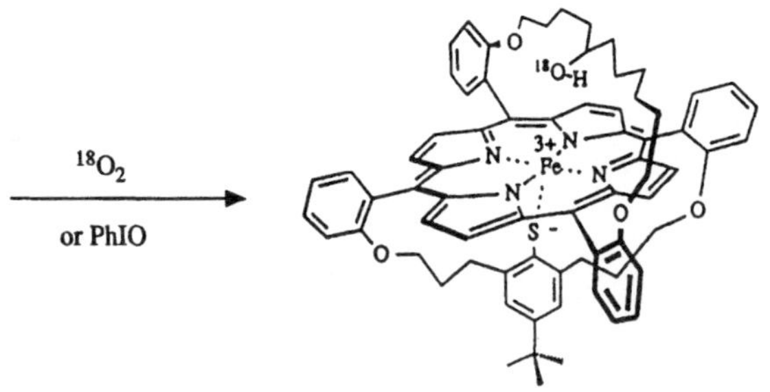

165

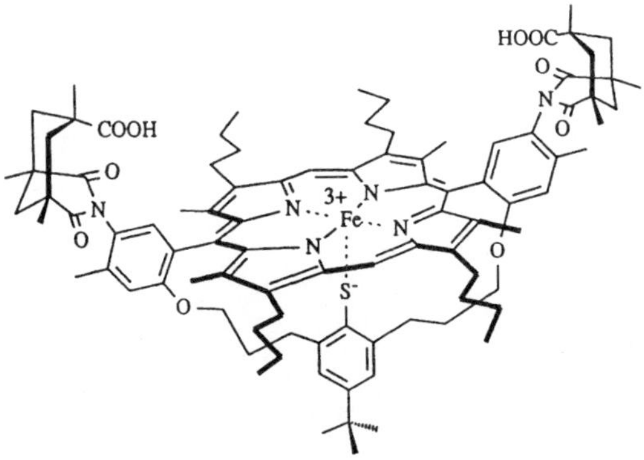

166

Fig. 27. An active site analogue of cytochrome P450 with substrate recognition sites and a thiophenolate ligand coordinating to the iron

shows a UV-/vis-spectrum similar to that of wild-type Mb (Soret band, $\lambda_{max} = 420$ nm). The reactivity of the iron(III)-MbHis(93)Cys towards O–O bond cleavage was investigated with cumene hydroperoxide, suggesting a preference of heterolytic over homolytic O–O bond scission. Using H_2O_2 the mutant displayed 5.1-fold increase of the initial reaction rate of epoxidation of styrene in comparison to wild-type Mb whereas the activity of N-demethylation of N,N-dimethylaniline was found to be about two-fold greater.

2.3 P450 Analogues with Substrate Recognition Sites

Beyond reactivity control the regio- and stereospecific insertion of "O" is another problem for the design of catalytic P450 models. Obviously, the only way to accomplish stereocontrol is binding the substrate reversibly in an enzyme like fashion through hydrogen bridges and/or hydrophobic interaction. This goal can be accomplished by attaching host molecules to the porphyrin such as cyclodextrins, crown ethers, and cyclophanes or creating recognition sites with the aid of Kemp's acids, BINAP residues, and appropriate bridges.

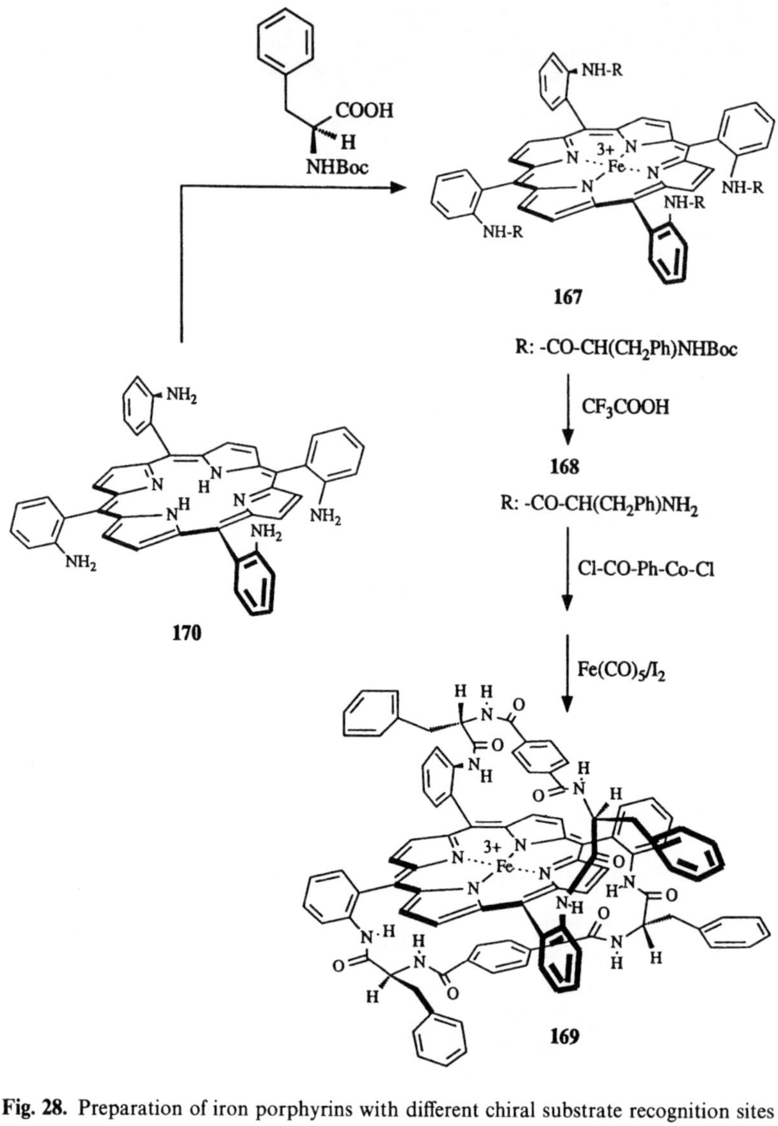

Fig. 28. Preparation of iron porphyrins with different chiral substrate recognition sites

From all this work only a few recent examples will be discussed for which a P450-like reactivity has been demonstrated.

In order to compare chiral recognition of the substrate *p*-chlorostyrene in the presence of chiral "*picket fences*" and "*basket handles*" carrying the same amino acids the porphyrins **167–169** were prepared in a straightforward manner from the known α,β,α,β-tetra(*o*-aminophenyl)porphyrin **170**, Fig. 28 [113]. Reaction with *t*-butoxycarbonyl-L-phenylalanine, removal of the protecting group and subsequent condensation with terephthaloyl chloride e.g. furnished the free base of **169**. Iron was always introduced using $Fe(CO)_5/I_2$ and the resulting metal complexes were employed in catalytic concentrations ($\leqslant 0.01\%$) in the presence of PhIO to epoxidize o-chlorostyrene.

Employing **167** and **168** the epoxides were obtained in yields $\leqslant 50\%$ displaying an excess of the (S)-enantiomer, 12% ee and 21% ee, respectively. When the more rigid "*basket handle*" porphyrin **169** was used the epoxide with (R)-configuration was found to be dominant (50% ee). In case of the "*basket handle*" porphyrin one could argue that the more pronounced enantioface selectivity is a consequence of a π-π interaction between the phenyl ring of the substrate and the phenyl group of the amino acid plus a small contribution of a hydrogen bridge between N–H$\cdots$Cl–Ph. In agreement with observation it would be expected that these effects are smaller in a less rigid environment like at the substrate binding site of **167** and **168** but one can neither predict nor interpret the preference for the opposite chirality.

Recently, two BINAP-protected iron porphyrins **171** [114] and **172** [115] have been prepared which are different with respect to the size of the hydrophobic cavity offered for substrate binding (Fig. 29).

The "eclipsed" tetra-BINAP porphyrin **171** was conveniently synthesized by condensation of the *meso*-tetrakis(2,6-dihydroxyphenyl)porphyrin **173** with the (S)-BINAP derivative **174** in the presence of K_2CO_3. After removal of the "staggered" isomer iron was inserted by addition of $Fe(CO)_5/I_2$, and the resulting Fe(III)-complex **171** was used as a catalyst (0.2%) to epoxidize a series of six styrene derivatives in the presence of an excess of PhIO. In every case the corresponding (R)-epoxides were preferentially formed in yields up to 72%. The best ee-values were obtained for the electron deficient substrates 2-nitrostyrene (80% ee) and pentafluorostyrene (74% ee), [114].

Both benzylic hydroxylations and olefin epoxidations were investigated with the "vaulted" Fe(III) and Mn(III) porphyrins **172** and **175**, easily prepared from $\alpha,\beta,\alpha,\beta$- *meso*-tetrakis(2-aminophenyl)porphyrin **170** and the (R)-dimethoxy-BINAP acid chloride **176**, followed by metal insertion using standard methods, [115]. The best results were obtained for the substrate (Z)-β-methylstyrene in the presence of 0.01% Fe(III) catalyst. The product mixture contained the (1S, 2R) enantiomer in 72% ee. However, it is obvious that the yields dropped sharply from 64% to 9% in attempts to raise the ee from 58% to the maximal 72%. It appears that the porphyrin **171** is slightly superior since under nearly identical conditions 80% ee were obtained with this catalyst in the epoxidation of 2-nitrostyrene along with a three times higher turnover number [114] (mol product/mol catalyst). For most olefins examined with the "vaulted" porphyrins the formation of the epoxide enantiomer produced in excess can be rationalized assuming that the larger olefin substituent occupies the open cleft of the spanning binaphthyl residue (Fig. 29) and that the olefin approaches the iron-oxo moiety in a side-on fashion.

Appreciably high ee-values were also observed in a series of hydroxylations at benzylic positions in all cases predominantly leading to the (R)-configured alcohol [115]. With tetrahydronaphthalene as substrate the best result was obtained: 72% ee, 47% yield.

A more detailed investigation using (R)-**177** and (S)-1-^{2}H$_1$-ethylbenzene **178** and the "vaulted" Fe(III) porphyrin **172** revealed an interesting enantioselectivity. It was discovered that the *pro*-R hydrogen was removed from the benzylic position preferentially to the *pro*-S by a factor of 2:1. Moreover, the radical formed by *pro*-R abstraction was trapped with ≥ 90% retention of configura-

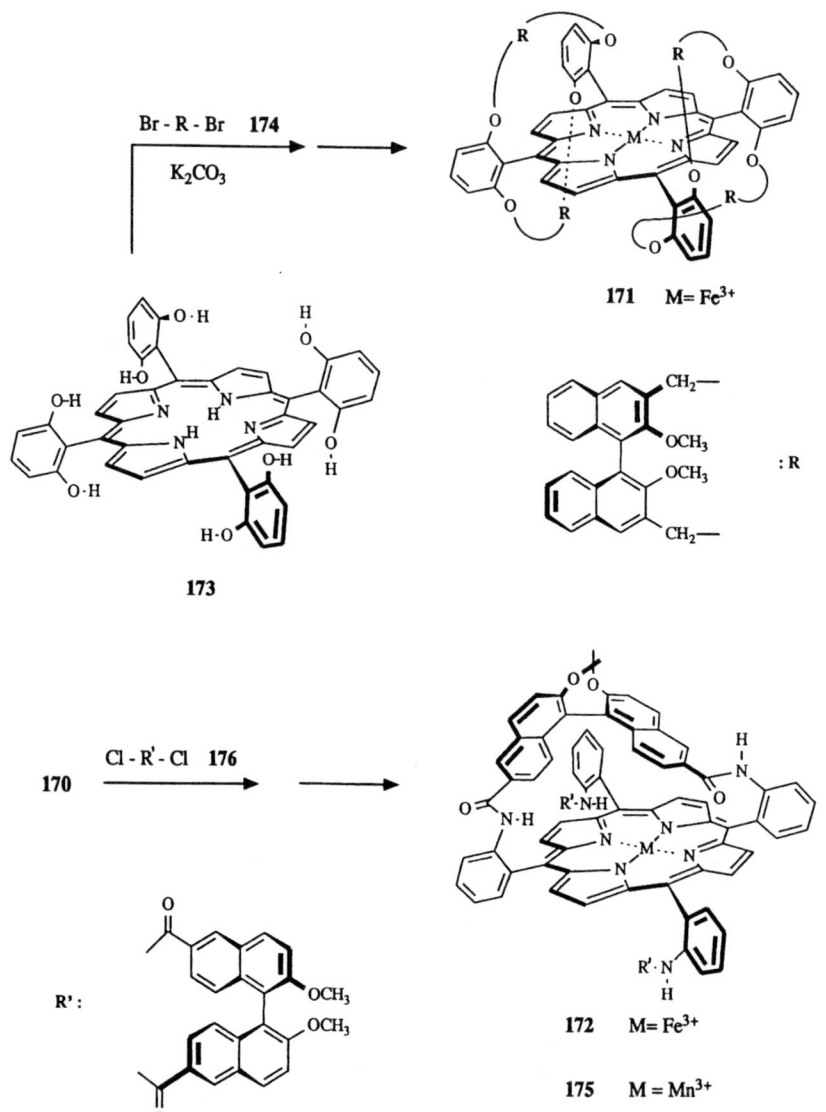

Fig. 29. Synthesis of binaphthyl bridged iron porphyrins and their application to benzylic hydroxylation

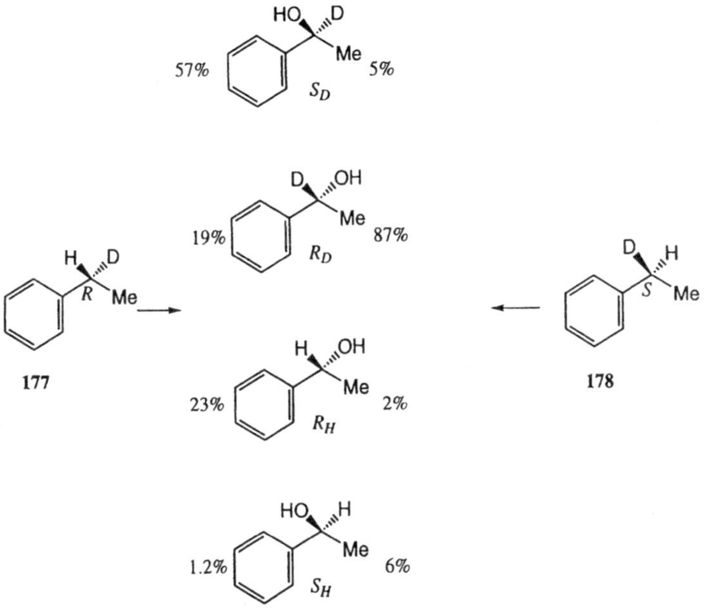

Fig. 29. Continued

tion; in contrast the radical produced by *pro*-S hydrogen removal gave up to 40% racemization (Fig. 29).

This result indicates that H· removal is less stereoselective than radical recombination and, moreover, the latter process seems to be kinetically different for both radicals due to different steric interactions with the rigid binaphthyl cavity. Epoxidations and hydroxylations were also carried out with the vaulted Mn(III) catalyst **175** under otherwise identical conditions but surprisingly the ee-values were much lower by a factor of 2 to 12.

Equally disappointing were the results obtained employing the Fe(III) catalyst **172** for the oxidation of prochiral sulfides [115]. Though the yields observed were up to 88% the maximal ee was only 48%. From these studies it appears that the more reactive substrates display the lowest ee, pointing to smaller energy differences in the diastereomeric transition states which are expected to be early on the reaction coordinate.

A very interesting molecular assembly has been designed recently which takes into account the often observed embedding of P450 enzymes into membranes. The general ideal was to prepare a porphyrin carrying hydrophobic orthogonal "appendices" on both faces so that the total length of the modified porphyrin would be complementary to phospholipid bilayers (35–40 Å). The porphyrin bilayer assembly is drawn in Fig. 30 [116].

For the synthesis of the porphyrin the formate ester of 3β-hydroxy-5-cholenic acid **179** was coupled via amide bonds to the α,β,α,β-atropisomer of *meso*-tetrakis(*o*-aminophenyl)-porphyrin **170**, using the mixed anhydride

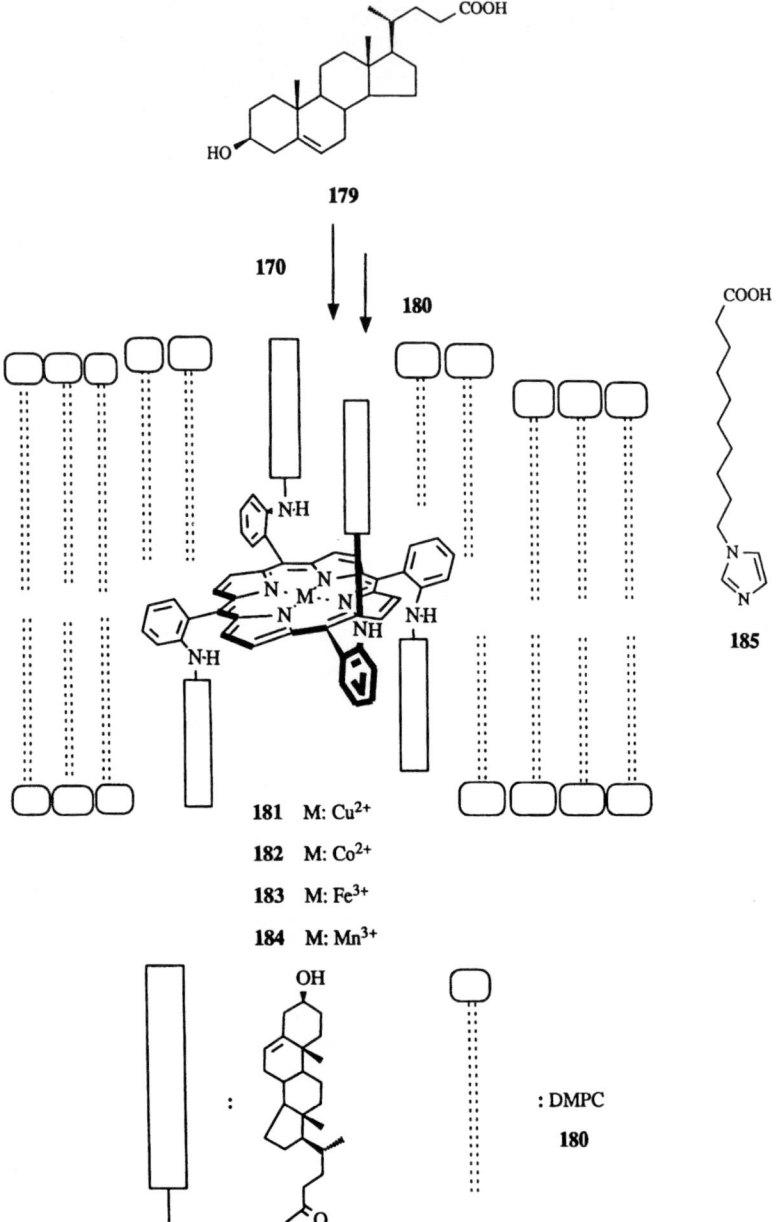

179

170

180

185

181 M: Cu²⁺
182 M: Co²⁺
183 M: Fe³⁺
184 M: Mn³⁺

: DMPC
180

Fig. 30. Metalporphyrins embedded in bilayers

method. After metallation unilamellar vesicles were prepared by ultrasonication of thin films of the porphyrin and DMPC **180** (dimyristoylphosphocholine) to yield the intercalated porphyrins **181–184**.

There are three lines of evidence for the incorporation of the porphyrin into the phospholipid bilayer: first, it was found that the water insoluble steroidal porphyrin coeluted with the vesicle on gel permeation chromatography; second, the existence of intact vesicles containing the Cu(II) porphyrin complex **181** was demonstrated by ^{31}P-NMR/Eu(NO$_3$)$_3$ spectroscopy giving rise to two singlets corresponding to the inner and the outer phosphocholine group of the vesicle; third, by using the ESR spectrum of the paramagnetic Cu(II) porphyrin **181** as a molecular probe in comparison to copper tetraphenylporphyrin crystals it became evident that the porphyrin ring is placed perpendicular to the bilayer normal as shown in Fig. 30. Furthermore, it was demonstrated that the porphyrin is centered between the two phospholipid layers with relatively small positional fluctuations ($\pm$ 3–4 Å) by using information obtained from the ESR spectra of the corresponding Co(II) porphyrin **182** in the presence of imidazole ligands **185** carrying variable hydrophobic chains. Only if $n \geqslant 6$, equal to the distance of 15–16 Å, was a change in ESR spectrum observed indicating that the imidazole coordinates to the metal.

With these highly oriented porphyrins regioselective oxidations were performed using the iron(III)- and manganese (III)-complexes **183** and **184**, respectively, in the presence of sodium periodate, PhIO or O$_2$/ascorbate.

The sterols that were chosen as substrates contained two double bonds, one at various positions in the side chain and Δ^5 in the steroid nucleus. Whereas the latter double bond was never touched in reactions with the Fe(III) porphyrin vesicle system **183** in the presence of PhIO, the side chain double bonds of desmosterol **186** and fucosterol **187** were epoxidized to **188** and **189** in 32% and 22% yield, respectively (Fig. 31). In contrast, stigmasterol **190** was not reactive, since the double bond cannot approach the reactive iron-oxo intermediate.

Polyunsaturated fatty acids were epoxidized less regioselectively, generally < 2:1 in favor of the double bond closer to the hydrophobic terminus. This result can be explained by a well known decrease of the rigidity of phospholipid bilayers on the addition of unsaturated fatty acids. However, when less than 20% cholesterol was added to the bilayer assembly the rigidity increased significantly and thus the regioselectivity.

Hydroxylations were performed with the steroidal Mn(III)porphyrin containing vesicles **184** in the presence of sodium periodate or under aerobic conditions using ascorbate as a reducing agent. The former more reactive conditions gave slightly better yields for the substrates cyclooctane and ethylbenzene. Using the latter conditions cholesterol **191** was exclusively hydroxylated at C-25, in agreement with the assumption that this C–H bond is closest to the active site (Fig. 32). But the yield of **192** was very low (2%); from the figures reported one can calculate that 80% of the catalyst was used up in the reaction and thus the system does not work catalytically. This hints towards a general problem of this very sophisticated system. It seems that substrates which are easily embedded into the phospholipid due to their hydrophobic nature will be oxidized by the metal porphyrin to yield products which are not sufficiently different in hydrophobicity to be released quickly from the vesicle.

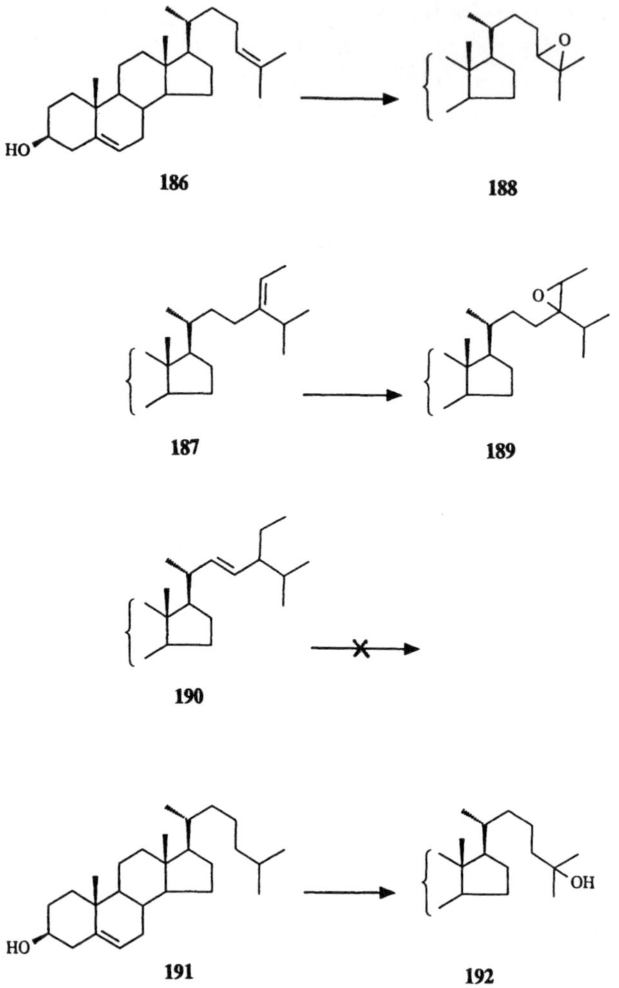

Fig. 31. Regio- and stereospecific oxidations with metalporphyrin vesicles

Moreover, products like alcohols will coordinate to the metal. Accordingly, product release may be rate determining for these "membrane spanning" steroidal metalloporphyrins.

Cyclophanes are known to be efficient receptors for aromatic compounds in protic solvents. Thus, linking a cyclophane unit to a porphyrin, like in **193**, provides an excellent way to study the oxidation of aromatic hydrocarbons [117]. The synthesis of **193** took advantage of an earlier protocol for the preparation of "strapped" porphyrins [118] using the bis-dipyrromethane **194** already linked to the cyclophane as a valuable precursor for an acid catalyzed condensation leading to the porphyrin in 9% yield (Fig. 32).

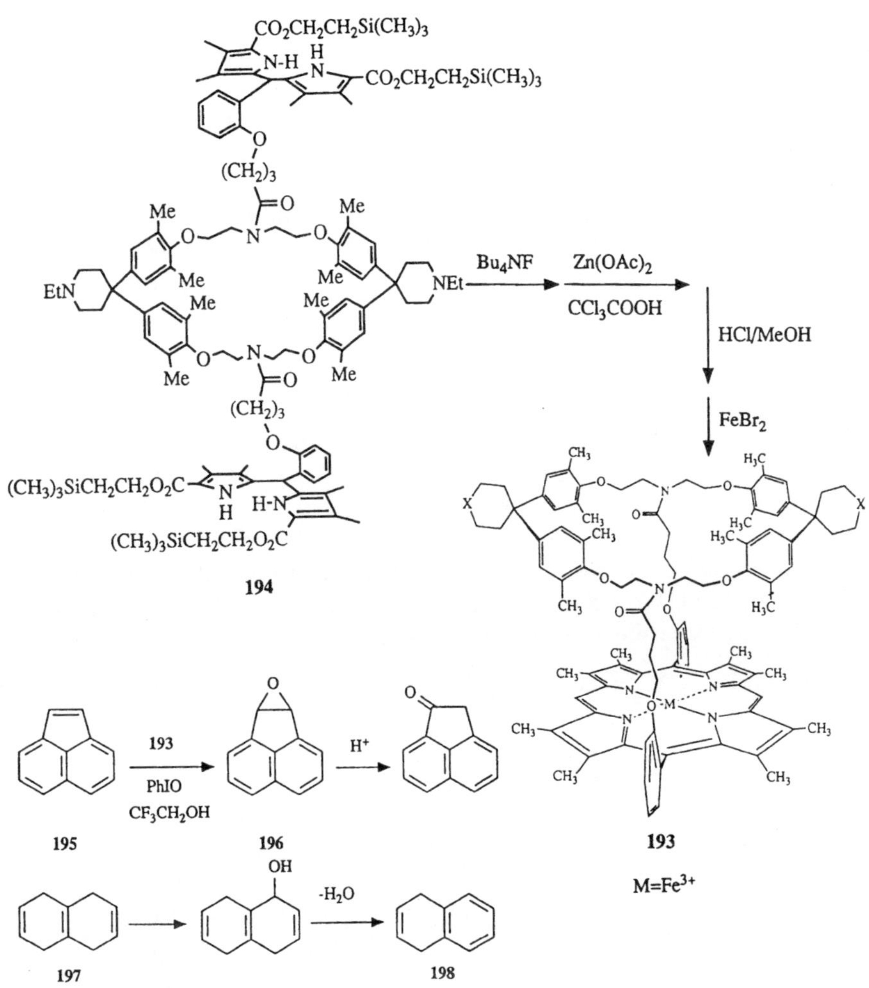

Fig. 32. Preparation of a cyclophane-bridged porphyrin

Several aromatic compounds were shown to bind reversibly in the cyclophane cavity of the free base, corresponding to **193**. For example, acenaphthylene **195** exposes its 1,2 double bond on top of the porphyrin plane and thus it binds in a productive orientation. In fact **195** was oxidized in the presence of 5 mol % of the high spin iron(III) complex **193** (PhIO/CF_3CH_2OH) initially to the epoxide **196** which finally rearranged to the corresponding ketone (28% yield, turnover 6). Another interesting reaction of **193** with isotetralin **197** seemingly involves allylic oxidation in the first place followed by removal of water to generate mainly 1,4-dihydronaphthalin **198**.

The asymmetric epoxidation of unfunctionlized olefins has been investigated with a new family of Mn(III)porphyrins which also involve the concept of "strapping" the porphyrin on one face and, moreover, shielding the opposite ligand sphere of Mn with a bulky imidazole ligand [119]. The synthesis of **199** was accomplished by condensation of the known tetrahydroxy porphyrin **153** with the threitol strap **200**. From the mixture of three isomers (46.3%) the *out/out*-isomer **201** (16%) was separated and its corresponding Mn(III) complex **202** was prepared. A systematic variation of imidazole ligands and different oxidants revealed that 1,5-dicyclohexylimidazole **203** and PhIO are the superior additives for the epoxidation of mono- and disubstituted (*Z*) olefins. The best results were obtained with e.g. **204**, 0.1% **202**, 10% PhIO and 25% **203** to furnish the (*1R, 2S*)-epoxide **205** with 88% ee (26% yield, based on PhIO consumption), and 84% ee (85% yield), respectively (Fig. 33). Although the ee values decrease significantly with increasing catalyst turnover numbers, the authors claim that their system is comparable to the well studied chiral manganese salen complexes [120].

The synthesis of the active site analogue **206** that combines two essential features of the P450 enzymes has recently been accomplished (Fig. 34) [111]. The design involves the attachment of the thiophenolate ligand to the porphyrin **207** prepared very efficiently (⩾ 80% yield) from the aldehyde **208** and the pyrromethane **209**. Condensation of the free amine **210** with the acid chloride of Kemp's acid **211** furnished **206** containing a substrate binding site at the porphyrin face opposite to the thiolate ligand. Preliminary experiments with this

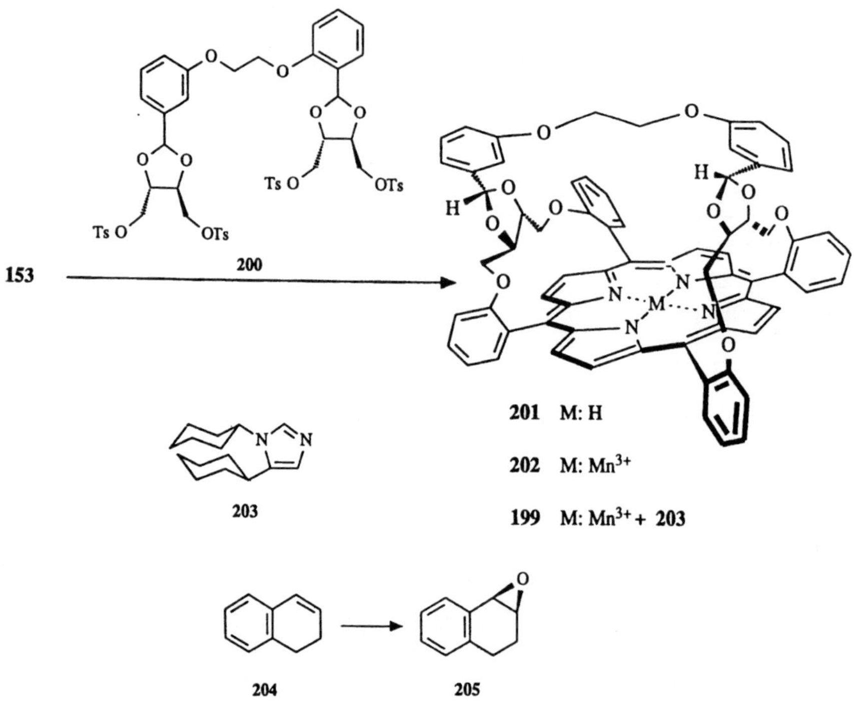

Fig. 33. Synthesis of threitol-strapped Mn^{3+} porphyrins as enantioselective catalysts

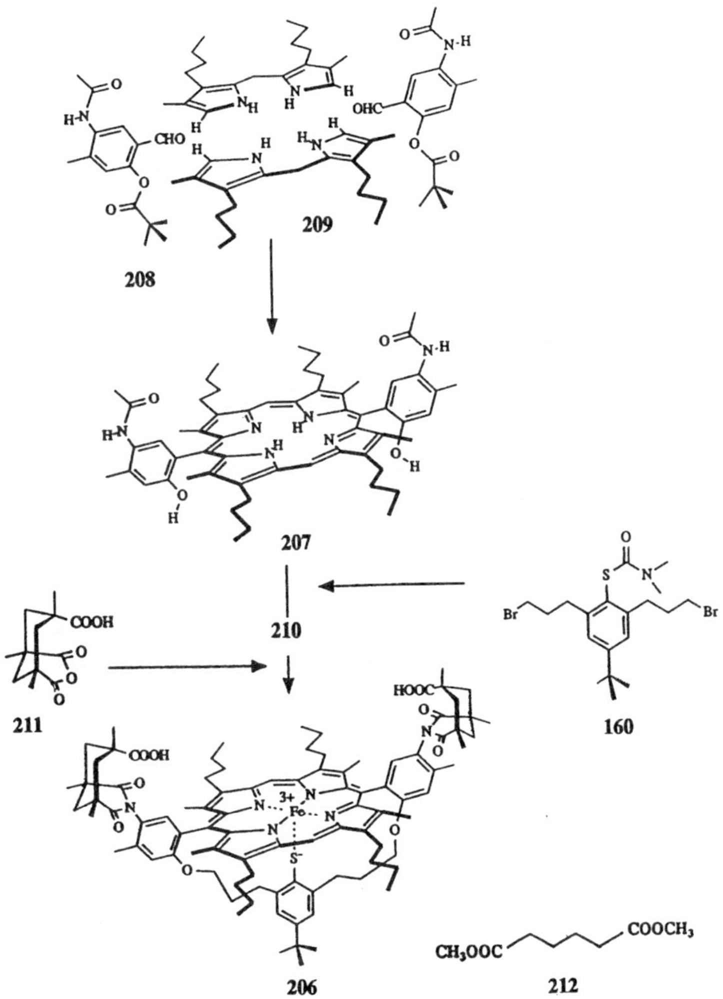

Fig. 34. Synthesis of an iron thiophenolate porphyrin with substrate binding sites

model compound revealed that polar solvent molecules, in particular water, are trapped in the cavity formed by the porphyrin plane and the Kemp acids thus mimicking the resting state of cytochrome P450. Moreover, **206** behaves spectroscopically and chemically like a P450 analogue displaying the characteristic shift of the *Soret* band on formation of the OC-iron(II)-S$^-$ and hydroxylation of substrates like **212** with a turnover of 150.

2.4 Electron Deficient Metal Porphyrins

With regard to preparing P450 models of practical value the reductive cleavage of molecular oxygen has always been regarded as the major obstacle. Accordingly, the discovery of a "shunt pathway" of the catalytic cycle using artificial "O" donors like PhIO and simple *meso*-tetraphenylporphyrinato-iron(III) complexes (Fe(TPP)X⁻) has been a considerable breakthrough in order to circumvent the problem [121, 122]. However, low turnover numbers due to irreversible oxidation of the metal porphyrin by the cooxidant was soon recognized as a serious complication.

In order to establish catalytic assemblies more resistant to autoxidation two strategies were invented: the careful selection of the cooxidant [123], and modification of the porphyrins by introducing halogen substituents peripherally. The latter approach is particularly intriguing because electron deficient porphyrins will also generate a more electrophilic and thus more reactive iron-oxene complex. As early as 1981 it was discovered [124] that pentafluorinated tetraphenyliron(III)porphyrin **213** (Fe(TF$_5$PP)Cl), Fig. 35, was significantly more stable than Fe(TPP)Cl **214** hydroxylating cyclohexane in 71% yield based on PhIO (turnover 7.1); in contrast **214** gave only 5% yield (turnover 0.5). **213** was also found to oxidize e.g. 4-^{2}H-anisol to *p*-methoxyphenol with 72% deuterium retention indicating a marked NIH shift characteristic of P450 catalyzed aromatic hydroxylations.

The hydroxylation of cyclohexane was also investigated using pentafluoroiodosobenzene (PhF$_5$IO) and the *meso*-tetra(2,6-dichlorophenyl)porphinatoiron(III)chloride **215** (Fe(TCl$_2$PP)Cl) which not only gave better turnovers, 45 ⇒ 440, at comparable yields but also seemed to be sufficiently sterically congested to resist μ-oxo dimer formation. An outstanding improvement was also achieved employing **215** *vs* **214** for the oxidation of norbornene: epoxinorbornene was isolated in 85% yield with a catalyst turnover of up to 10 000 [125]. Even more resistance towards oxidative degradation of the chromophore was obtained replacing the β-pyrrol-H for halogen substituents. The treatment of Zn(TCl$_2$PP)Cl with a tenfold excess of NBS in CCl$_4$ was reported to yield an octabromoderivative in 71% yield, which after removal of Zn with CF$_3$COOH and subsequent insertion of iron (FeBr$_2$/DMF) led to the isolation of octabromoiron(III)porphyrin **216** (Fe(TCl$_2$PBr$_8$P)Cl) [126]. The bromination procedure was recently reinvestigated and found not to work well neither in CCl$_4$ nor in CHCl$_3$, only in C$_2$Cl$_4$ could chlorination of the porphyrins periphery be suppressed and pure (Zn(TCl$_2$PBr$_8$P)) was obtained in 80% yield [127]. Alternatively, it was shown that zinc tetramesityl porphyrins can be readily brominated with NBS in methanol and that the corresponding Mn(III)- and Fe(III)-complexes **217** are efficient catalysts for the oxidation of lignin analogues [128]. The most stable iron(III)porphyrin which can be used even in the presence of H$_2$O$_2$ was obtained by fluorinating a Zn(TF$_5$PP) with e.g. cobalt fluoride in CH$_2$Cl$_2$/pyridine from which the corresponding iron(III)porphyrin o**218** was prepared by means of known metal exchange reactions [129].

Accordingly, the consecutive introduction of electron withdrawing groups in Fe(TPP)Cl **214** ⇒ Fe(TF$_5$PF$_8$P)Cl **218** has a dramatic effect with respect to the stability both towards oxidative degradation and μ-oxo dimer formation. This

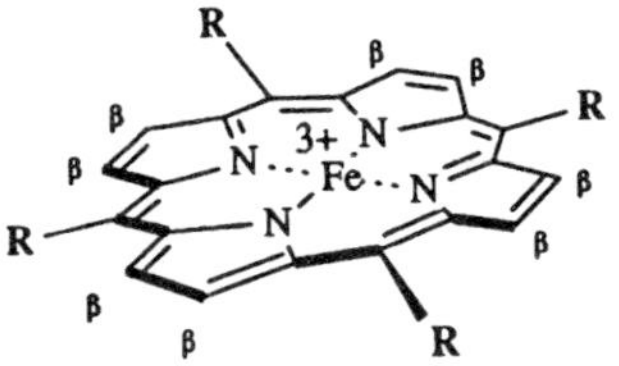

213 β = H **R** : 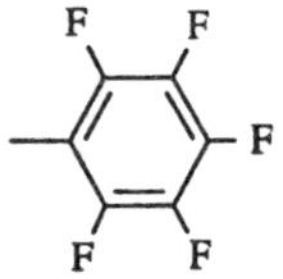

217 β = Br **R** : 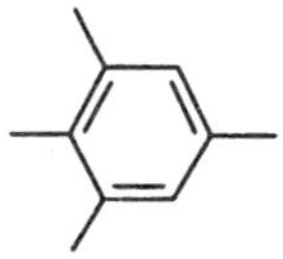

214 β = H; R : Ph

215 β = H

216 β = Br

} **R** : 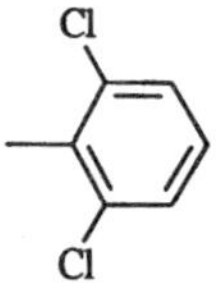

218 β = F

219 β = Cl

229 β = Br

} **R** : 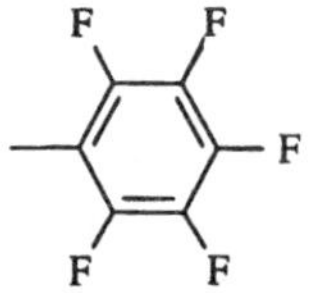

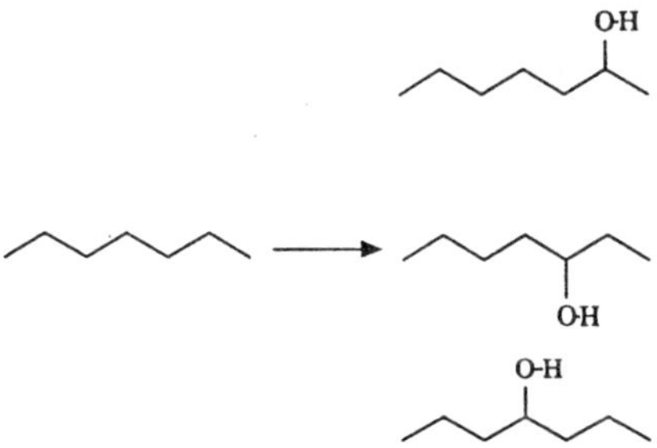

213	**215**	**216**
40%	58%	42%
40%	30%	45%
20%	12%	16%

Fig. 35. Regioselective hydroxylations with perhalogenated metalporphyrins

was interpreted in terms of increasing electron deficiency of the macrocyclic ligand rather than as a consequence of steric hindrance [122].

In an attempt to compare the intrinsic reactivities of certain perhalogenated iron porphyrins the hydroxylation of heptane was investigated in the presence of PhIO [130]. Among the catalysts having no halogen substituents at the β-pyrrol positions, only the pentafluorinated Fe(III)porphyrin **213** gave good yields (total 70%) of heptanols (Fig. 35). A remarkable increase of the yields from 38% to 80% was observed employing **215** and **216**, respectively. Concerning the regioselectivity of hydroxylation, the terminal methyl groups were poorly attacked and, with sterically non demanding porphyrins, the secondary alcohols were formed in an almost statistical ratio (40:40:20). However, in case

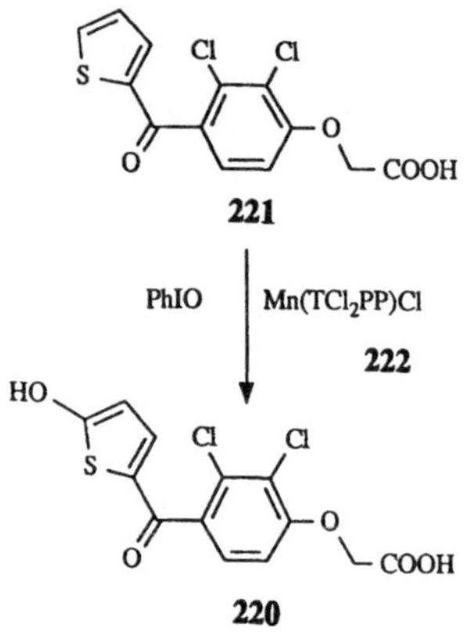

Fig. 36. Hydroxylation of tienilic acid

of the more bulky iron(III)porphyrin **215** hydroxylation at the more accessible C(2) was favored (58:30:12). When halogen substituents were placed at the periphery, e.g. **216**, surprisingly this trend was reversed to a statistic distribution. The effect can be understood assuming a change of the electronic character of the reactive iron-oxene species: In the presence of electronegative substituents an enhanced reactivity is observed due to the tendency of the radical resting on the oxygen rather than on the porphyrin. This hypothesis is supported by experiments comparing the chemoselectivity of oxidation catalyzed by the porphyrins with a mixture of cyclooctene/heptane 1:100. Each secondary C–H bond of heptane is about 1500 times less reactive than the cyclooctene double bond. With electron rich porphyrins like *meso*-tetramesityliron(III)porphyrin the ratio of epoxide/heptanols was about 10:1, but for the perhalogenated **219** the ratio was only 1.5:1 indicating an enhanced reactivity/chemoselectivity for C–H bond hydroxylation.

Progress has been reported concerning the application of perhalogenated porphyrins to mimic P450 catalyzed drug metabolism. For example 5-hydroxytielinic acid **220**, the main *in vivo*- and *in vitro*-metabolite of tielinic acid **221**, can be prepared in good yields using PhIO and *meso*-tetrakis(2,6-dichlorophenyl)Mn(III)porphyrin **222**, Fig. 36 [121]. The mechanism of this transformation most likely proceeds via the sulfoxide which undergoes a rearrangement reminiscent of the Pummerer reaction to furnish the hydroxylated thiophene **220**.

2.5 Reductive Oxygen Cleavage

The bond dissociation energies of O=O (118 kcal/mol) and related species like HO–O˙ (65 kcal/mol) and –O–O⁻ (~50 kcal/mol) predict that the O–O bond is weakened by the addition of electrons and protons. This is confirmed by kinetic studies demonstrating that O–O bond scission is accelerated 4×10^6-fold by

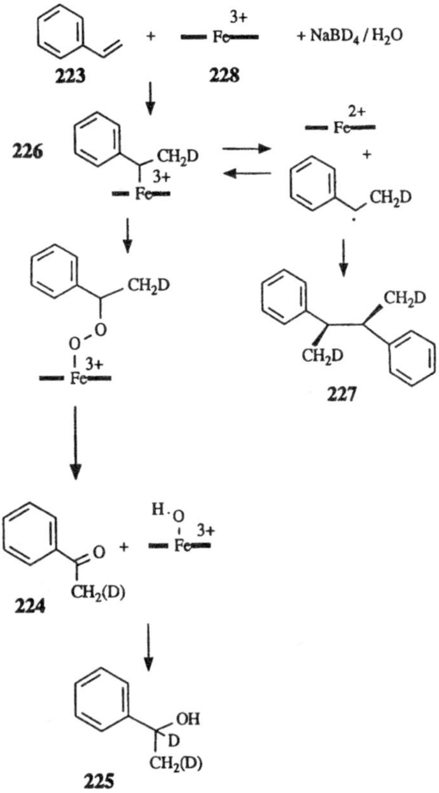

Fig. 37. Unusual hydroxylation with iron(III)TMPyP, O_2, and NaBD$_4$

electron transfer to oxygen complexes of iron(II)-"*picket fence*"-porphyrins [131].

NaBH$_4$ has been used as an efficient electron donor for reductive oxygen fission in the presence of Mn(III)porphyrins but it was recognized that in the presence of metalporphyrins BH$_4^-$ readily cleaves epoxides to alcohols and reduces ketones [132]. Detailed recent studies suggest that the primary product of the oxidation of styrene **223** is acetophenone **224**, which subsequently is reduced to 1-phenylethanol **225**. The mechanism does not involve a reactive iron-oxene intermediate but a σ-alkyl complex **226** which was postulated because 1,4-dideuterio-2,3-diphenylbutane **227** could be isolated at 2% by performing the reaction with NaBD$_4$ and *meso*-tetrakis(1-methyl-4-pyridino) iron(III)porphyrin **228** (FeTMPyP), Fig. 37. A similar result was obtained with the system Mn(III)tetraphenylporphyrin/O$_2$/NaBH$_4$/methanol but in the presence of *N*-methylimidazole as the fifth ligand [133]. High conversion rates (up to 100%) of the olefin were observed with an excess of NaBH$_4$, although it was found that bulky substituents at the double bond increase the formation of sideproducts.

The oxidation of cyclohexene was systematically investigated by means of metal porphyrins which have distinct redox potentials due to different metals [134]. The best results were obtained with Mn(III)TPP/L-Cys/NaBH$_4$ which furnished a product mixture consisting of cyclohexenone (46.4%) > cyclohexanol (23.8%) > cyclohexenol (19.5%) > cyclohexanone (9.0%) > epoxide (1.4%), relative yields given in parenthesis. In the presence of KOH and riboflavin as an electron transfer reagent the product distribution was similar but the total yield was considerably improved [135].

On adding a "promoter" like microcrystalline cellulose (Avicel^R) to the system Mn(III)TPP/NaBH$_4$/O$_2$/benzene cyclohexcne was converted mainly to cyclohexanol (51.7%) and cyclohexenol (45.3%) with a total yield of 17.23% based on olefin (turnover 1994/h) [136].

The product mixtures obtained using the various NaBH$_4$-containing assemblies often resemble the distribution obtainable with hepatic microsomes, although synthetic applications seem to be very limited. In this context reductive oxygen activation through use of H$_2$/colloidal Pt/Mn(III)-"*picket fence*"-TPP/N-methylimidazole offers a clear advantage [131]. Though the turnovers (< 100) are generally smaller than for perhalogenated Mn(III)porphyrins, the regio- and stereospecificity of e.g. epoxidations are remarkable. Diolefins like geranylacetate are oxidized in favor of the terminal double bond 97:3. Moreover, this catalyst performs the whole spectrum of P450 enzyme catalyzed reactions like aromatic hydroxylation, N-dealkylation and aliphatic hydroxylation. The latter proceeds slowly but in favor of tertiary hydroxylation.

With respect to turnover numbers and yields significant improvement has been accomplished using Zn powder as the reducing agent and acetic acid as the proton source to produce the "O"-transfer intermediate (Mn(V)=O) from Mn(III)TPP/N-methylimidazole [137]. With this system a number of substrates have been shown to be oxidized in good yields (15–70%/turnover 30–180) even on a gram scale, and some examples are given in Fig. 38. This method seems to be favored over an earlier two-phase system using sodium ascorbate as a reducing agent [138] and also superior to Zn amalgam in the presence of iron(III) porphyrins/methyl viologen and stoichiometric amounts of acetic anhydride [139].

A quite unusual oxygen cleavage has been observed in the absence of any reducing agent employing the completely halogenated porphyrin Fe(TF$_5$TBr$_8$P) Cl **229**, see Fig. 35. This catalyst oxidized isobutane with > 90% selectivity to afford *tert*-butanol with a turnover of 13 560, although under strict safety conditions at 25 °C [140]. It was suggested that the electron withdrawing groups at the periphery of the porphyrin prevent the formation of μ-oxodimers from Fe(IV) = O which seems to be the reactive species performing oxygen insertion.

3 Final Comment

This account has been written to make the non-expert aware of one of the most important enzyme families in organisms living under aerobic conditions.

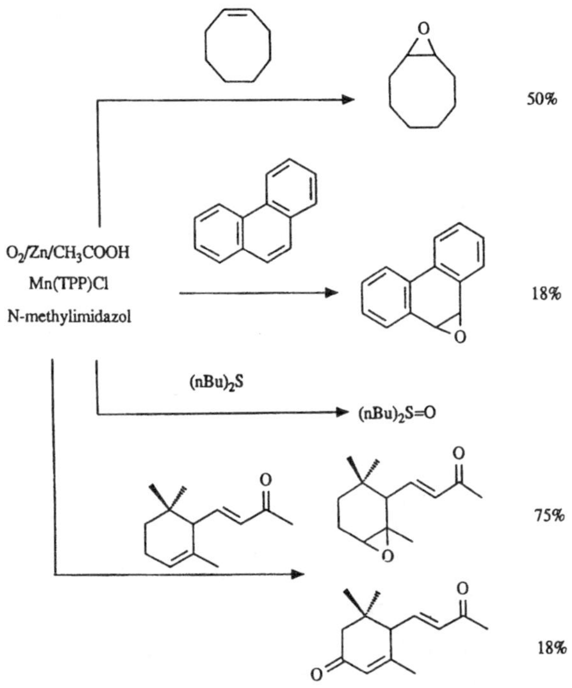

O_2/Zn/CH_3COOH
Mn(TPP)Cl
N-methylimidazol

(nBu)_2S

(nBu)_2S=O

50%

18%

75%

18%

Fig. 38. Various epoxidations using Mn(III)TPP Cl/O_2 under reductive conditions

Limited space and time imposed restrictions to report only a few out of many experiments in each area of P450 research which were selected either for their conclusiveness or for their controversial nature. Accordingly, this overview has a personal attitude which is also obvious concerning the following two aspects. Often I have interpreted results in view of investigations at about the same time or thereafter and thus the literature cited may neither include these arguments nor may the authors agree. Furthermore, I have applied some consistency in drawing electronic structures and writing formal charges which, though they often do not appear in the original literature, will hopefully make reading easier. Those who need further informations on cytochrome P450 or other heme-thiolate proteins are advised to consult references [141, 142 and 143].

4 References

1. Omura T, Sato R (1964) J Biol Chem 239: 2370
2. Sligar SG, Murray RI (1986) In: Ortiz de Montellano PR (ed) Cytochrome P450, structure, mechanism and biochemistry, Plenum, New York, p 429
3. Durst F (1991) In: Ruckpaul K, Rein H (eds) Frontiers in biotransformation, Akademie, Berlin, p 191
4. Guengrich FP (1987) Mammalian cytochromes P450, CRC, Boca Raton, Florida, vol 1–2

5. Ortiz de Montellano PR (ed) (1986) Cytochrome P450, structure, mechanism and biochemistry, Plenum, New York
6. Stern JO, Peisach J (1974) J Biol Chem 249: 7495
7. Collman JP, Sorrell TN, Hoffman BM (1975) J Am Chem Soc 97: 913
8. Chang CK, Dolphin D (1975) J Am Chem Soc 97: 5948
9. Loew GH, Herman ZS, Rohmer M-M, Goldblum A, Pudzianowsky A (1981) Ann NY Acad Sci 367: 192
10. Dawson JH (1988) Science 240: 433
11. Woggon W-D (1988) Nachr Chem Tech Lab 36: 890
12. Ravichandran KG, Boddupalli SS, Hasemann CA, Peterson JA, Deisenhofer J (1993) Science 261: 731
13. Poulos TL, Finzel BC, Howard AJ (1987) J Mol Biol 195: 687
14. Hasemann CA, Ravichandran KG, Peterson JA, Deisenhofer J (1994) J Mol Biol 236: 1169
15. Poulos TL, Finzel BC, Howard AJ (1986) Biochemistry 25: 5314
16. Raag R, Poulos TL (1991) Biochemistry 30: 2674
17. Raag R, Poulos TL (1989) Biochemistry 28: 7586
18. Tsai R, Yu CA, Gunsalus IC, Peisach J, Blumberg W, Orme-Johnson WH, Beinert H (1970) Proc Nat Acad Sci USA 66: 1157
19. Raag R, Poulos TL (1992) In: Ruckpaul K, Rein H (Eds) Frontiers in Biotransformation, Akademie Verlag Berlin, vol 7, p 1
20. Harris D, Loew GH (1993) J Am Chem Soc 115: 8775
21. Fisher MT, Sligar SG (1985) J Am Chem Soc 107: 5018
22. Sligar SG, Gunsalus IC (1976) Proc Natl Acad Sci USA 73: 1078
23. Schappacher M, Ricard L, Fischer J, Weiss R, Bill E, Montiel-Montoya R, Winkler H, Trautwein AX (1987) Eur J Biochem 168: 419
24. Gerothanassis IP, Momenteau M, Loock B (1989) J Am Chem Soc 111: 7006
25. Hintz MJ, Mock DM, Peterson LL, Tuttle K, Peterson JA (1982) J Biol Chem 257: 14324; Brewer CB, Peterson JA (1988) J Biol Chem 263: 791; Tyson CA, Lipscomb JD, Gunsalus IC (1972) J Biol Chem 247: 5777
26. Sligar SG, Kennedy KA, Pearson DC (1980) Proc Nat Acad Sci USA 77: 1240
27. Groves JT, Watanabe YJ (1988) J Chem Soc 110: 8443
28. Patzelt H, Woggon W-D (1992) Helv Chim Acta 75: 523
29. White RE, Coon MJ (1982) J Biol Chem 257: 3073
30. Traylor TG, Xu F (1990) J Am Chem Sic 112: 178
31. Imai Y, Shimada H, Watanabe Y, Matsushima-Hibiya Y, Makino R, Koga H, Horiuchi T, Ishimura Y (1989) Proc Nat Acad Sci USA 86: 7823
32. Gerber NC, Sligar SG (1992) J Am Chem Soc 114: 8742; Martinis SA, Atkins WM, Stayton PS, Sligar SG (1989) J Am Chem Soc 111: 9252
33. Groves JT, McClusky GA, White RE, Coon MJ (1978) Biochem Biophys Res Commun 81: 154
34. Keinan E, Mazur Y (1976) Synthesis 523; Cohen Z, Keinan E, Mazur Y, Varkony TH (1975) J Org Chem 40: 2141
35. Adam W, Curci R, Edwards JO (1989) Acc Chem Res 22: 205
36. DesMarteau DD, Donadelli A, Montanari V, Petrov VA, Resnati G (1993) J Am Chem Soc 115: 4897
37. Barton DHR, Doller D (1992) Acc Chem Res 25: 504
38. Shapiro S, Piper JU, Caspi E (1982) J Am Chem Soc 104: 2301
39. Jones JP, Rettie AR, Trager WF (1990) J Med Chem 33: 1242
40. Jones JP, Korzekwa KR, Rettie AR, Trager WF (1986) J Am Chem Soc 108: 7074; Jones JP, Trager WF (1987) J Am Chem Soc 109: 2171
41. Northrop DB (1975) Biochemistry 14: 2644
42. Ortiz de Montellano PR, Stearns RA (1987) J Am Chem Soc 109: 3415
43. Ortiz de Montellano PR, Stearns RA (1989) Drug Metabolism Reviews 20: 183
44. Bowry VW, Ingold KU (1991) J Am Chem Soc 113: 5699
45. Atkins WA, Sligar SG (1989) J Am Chem Soc 111: 2715
46. Gelb MH, Heimbrook DC, Malkonen P, Sligar SG (1982) Biochemistry 21: 370
47. Blake CR, Coon MJ (1981) J Biol Chem 256: 12127; White RE, Sligar SG, Coon MJ (1980) J Biol Chem 255: 11108
48. Gelb MH, Malkonen P, Sligar SG (1982) Biochem Biophys Res Commun 104: 853
49. Eble KS, Dawson JH (1984) J Biol Chem 259: 14389

50. Yoshida Y, Aoyama Y (1991) In: Ruckpaul K, Rein H (Eds) Frontiers in Biotransformation, Akademie Verlag Berlin, Vol 4, p 127
51. Byon CY, Gut M (1980) Biochem Biophys Res Commun 94: 549
52. Corina DL, Miller SL, Wright JN, Akhtar M (1991) J Chem Soc Chem Commun 782
53. Akhtar M, Njar VCO, Wright JN (1993) J Steroid Biochem Molec Biol 44: 375; Wright JN, Akhtar M (1990) Steroids 55: 142
54. Caspi E, Arunachalam T, Nelson PA (1986) J Am Chem Soc 108: 1847
55. Stevenson DE, Wright Jn, Akhtar M (1988) J Chem Soc Perkin Trans I 2043
56. Beusen DD, Carrell HL, Covey DF (1987) Biochemistry 26: 7833
57. Caspi E, Wicha J, Arunachalam T, Nelson P, Spiteller G (1984) J Am Chem Soc 106: 7282
58. Fishman J, Raju MS (1981) J Biol Chem 256 4472
59. Watanabe Y, Ishimura Y (1989) J Am Chem Soc 111: 8047; idem ibid 410
60. Vaz ADN, Roberts ES, Coon MJ (1991) J Am Chem Soc 113: 5886
61. Korzekwa KR, Trager WF, Smith SJ, Osawa Y, Gilette JR (1991) Biochemistry 30: 6155
62. Burstein S, Middleditch BS, Gut M (1975) J Biol Chem 250: 9028
63. Groves JT, Subramanian DV (1984) J Am Chem Soc 106: 2177
64. McClanahan RH, Huitric AC, Pearson PG, Desper JC, Nelson SD (1988) J Am Chem Soc 110: 1979
65. Fretz H, Woggon W-D (1986) Helv Chim Acta 69: 1959
66. Fretz H, Woggon W-D, Voges R (1989) Helv Chim Acta 72: 391
67. Floss HG, Tsai MD (1979) In: Meister A (Ed) Advances in Enzymology and Related Areas of Molecular Biology, Wiley & Sons, New York, p 253
68. Ortiz de Montellano PR, Mangold BLK, Wheeler C, Kunze KL, Reich NO (1983) J Biol Chem 258: 4208
69. Ostovic D, Bruice TC (1992) Acc Chem Res 25: 314
70. Collman JP, Hampton PD, Braumann JI (1990) J Am Chem Soc 112: 2986; Groves JT, Nemo TE (1983) J Am Chem Soc 105: 5786
71. Traylor TG, Nakano T, Dunlap BE, Traylor PS, Dolphin D (1986) J Am Chem Soc 108: 2782
72. Loew G, Collins J (1992) In: Ruckpaul K, Rein H (Eds) Frontiers in Biotransformation, Akademie Verlag, Berlin, Vol 7, p 90
73. Ortiz de Montellano PR, Fruetel JA, Collins JR, Camper DL, Loew G (1991) J Am Chem Soc 113: 3195
74. Kunze KL, Mangold BLK, Wheeler C, Beilan HS, Ortiz de Montellano PR (1983) J Biol Chem 258: 4202
75. Collman JP, Hampton PD, Brauman JI (1990) J Am Chem Soc 112: 2977
76. Groves JT, Avaria Neisser GE, Fish KMImachi M, Kuczkowski RL (1986) J Am Chem Soc 108: 3837
77. White RE, Miller JP, Favreau LV, Bhattacharyya A (1986) J Am Chem Soc 108: 6024
78. Groves JT, Viski P (1989) J Am Chem Soc 111: 8537
79. Hanzlik RP, Ling K-H (1990) J Org Chem 55: 3992
80. Hanzlik RP, Schafer AR, Moon JB, Judson CM (1987) J Am Chem Soc 109: 4926
81. Lewis DVW (1992) In: Ruckpaul K, Rein H (Eds) Frontiers in Biotransformation, Akademie Verlag Berlin, Vol 7, p 90
82. Tomaszewski JE, Jerina DM, Daly JW (1975) Biochemistry 14: 2024
83. Korzekwa K, Trager WF, Gouterman M, Spangler D, Loew G (1985) J Am Chem Soc 107: 4237
84. Korzekwa K, Swinney DC, Trager WF (1989) Biochemistry 28: 9019
85. Hanzlik RP, Hogberg K, Judson CM (1984) Biochemistry 23: 3048
86. Ortiz de Montellano PR, DeVoss J, Fruetel JA, Mackman RL, Peterson JA (1994) at the 10th International Symp. on Microsomes and Drug Oxidation, Toronto
87. Harada N, Miwa GT, Walsh JS, Lu AYH (1984) J Biol Chem 259: 3005
88. Tullman RH, Walsh JS, Miwa GT (1984) Fed Proc 43: 346
89. Guengrich FP (1987) J Biol Chem 262: 8459
90. Miwa GT, Walsh JS, Kedderis GL, Hollenberg PF (1983) J Biol Chem 258: 14445
91. Shono T, Toda T, Oshino N (1982) J Am Chem Soc 104: 2639
92. Shea JP, Valentine GL, Nelson SD (1982) Biochem Biophys Res Commun 109: 231
93. Kedderis GL, Dwyer LA, Rickert DE, Holenberg P (1983) Mol Pharmacol 23: 758
94. Hall LR, Hanzlik RP (1990) J Biol Chem 265: 12349
95. Hall LR, Iwamoto RT, Hanzlik RP (1989) J Org Chem 54: 2446

96. Augusto O, Beilan HS, Ortiz de Montellano PR (1982) J Biol Chem 257: 11288
97. Boucher JL, Genet A, Vadon S, Delaforge M, Mansuy D (1992) Biochem Biophys Res Commun 184: 1158; idem ibid 187: 880
98. Feldman PL, Griffith OW, Stuehr DJ (1993) CEN 26
99. Andronik-Lion V, Boucher JL, Delaforge M, Henry Y, Mansuy D (1992) Biochem Biophys Res Commun 185: 452
100. Feelisch M, Stamler JS (Eds) (1996) Methods in nitric oxide research, John Wiley, New York
101. Meunier B (1992) Chem Rev 92: 1411
102. Baldwin JE, Permutter P (1984) Top Curr Chem 121: 181
103. Battersby AR, Howson W, Hamilton AD (1982) J Chem Soc Chem Commun 1266
104. Collman JP, Brauman JI, Doxsee KM, Halbert TR, Bunnenberg E, Linder RE, LaMar GN, Del Gaudio J, Lang G, Spartalian K (1980) J Am Chem Soc 102: 4182
105. Stäubli B, Fretz H, Piantini U, Woggon WD (1987) Helv Chim Acta 70: 1173
106. Momenteau M, Mispelter J, Loock B, Lhoste JM (1985) J Chem Soc Perkin Trans 1: 61
107. Tatsuno Y, Tomita K, Tani K (1988) Inorganica Chimica Acta 152: 5
108. Higuchi T, Uzu S, Hirobe M (1990) J Am Chem Soc 112: 7051
109. Collman JP, Groh S (1982) J Am Chem Soc 104: 1391
110. Woggon WD, Matile S, Stäubli B in preparation
111. Ghirlanda S, Gmür C, Woggon WD in preparation
112. Adachi SI, Nagano S, Ishimori K, Watanabe Y, Morishima I (1993) Biochemistry 32: 241
113. Mansuy D, Battioni P, Renaud JP, Guerin P (1985) J Chem Soc Chem Commun 155
114. Naruta Y, Tani F, Maruyama K (1989) Chem Letters 1269
115. Groves JT, Viski P (1990) J Org Chem 55: 3628
116. Groves JT, Neumann R (1987) J Am Chem Soc 109: 5045; Groves JT, Neumann R (1988) J Org Chem 53: 3891; Groves JT, Neumann R (1989) J Am Chem Soc 111: 2900
117. Benson DR, Valentekovich R, Tam SW, Diederich F (1993) Helv Chim Acta 76: 2034
118. Baldwin JE, Crossley MJ, Klose T, O'Rear EA (1982) Tetrahedron 38: 27
119. Collman JP, Lee VJ, Kellen-Yuen CJ, Zhang X, Ibers JA, Brauman JI (1995) J Am Chem Soc 117: 692
120. Jacobsen EN, Zhang X, Muci AR, Ecker JR, Deng L (1991) J Am Chem Soc 113: 7063
121. Mansuy D, Battioni P, Battioni JP (1989) Eur J Biochem 184: 267
122. Traylor TG (1991) Pure & Appl Chem 63: 265
123. Collman JP, Tanaka H, Hembre RT, Brauman JI (1990) J Am Chem Soc 112: 3689
124. Chang CK, Ebina F (1981) J Chem Soc Chem Commun 778
125. Traylor PS, Dolphin D, Traylor TG (1984) J Chem Soc Chem Commun 279
126. Traylor TG, Tsuchiya S (1987) Inorg Chem 26: 1338
127. Gonzalves AM, Johnstone RAW, Pereira MM, Shaw J, Sobral AJF do N (1991) Tetrahedron Lett 32: 1355
128. Labat G, Meunier B (1989) J Org Chem 54: 5008; Hoffmann P, Labat G, Robert A, Meunier M (1990) Tetrahedron Lett 31: 1991
129. Tsuchiya S, Seno M (1989) Chemistry Lett 263
130. Bartoli JR, Brigaud O, Battioni P, Mansuy D (1991) J Chem Soc Chem Commun 440
131. Tabushi I (1988) Coordination Chem Rev 86: 1
132. Kano K, Takagi H, Takeuchi M, Hashimoto S, Yoshida Z (1991) Chemistry Lett 31: 519
133. Shimizu M, Orita H, Hayakawa T, Takehira K (1989) J Mol Cat 53: 165
134. Sakurai H (1988) J Mol Cat 47: 1
135. Sakurai H, Mori Y, Shibuya M (1989) Inorg Chim Acta 162: 23
136. Nishiki M, Satoh T, Sakurai H (1990) J Mol Cat 62: 79
137. Mansuy D (1990) Pure & Appl Chem 62: 741
138. Fontecave M, Mansuy D (1984) Tetrahedron 40: 4297
139. Karasevich EI, Khenkin AM, Shilov AE (1987) J Chem Soc Chem Commun 731
140. Lyons JE, Ellis Jr PE (1991) Catalysis Lett 8: 45
141. Gunter MJ, Turner P (1991) Coordination Chemistry Reviews 108: 115
142. Watanabe Y, Groves JT (1992) The Enzymes 20: 405
143. Ortiz de Montellano PR (Ed) (1995) Cytochrome P450. Structure, Mechanism and Biochemistry 2nd edn. Plenum Press, New York.

Enzymatic C–C Bond Formation in Asymmetric Synthesis

Wolf-Dieter Fessner and Christiane Walter

Institut für Organische Chemie der Rheinisch-Westfälischen Technischen Hochschule Aachen, Professor-Pirlet-Str. 1, D-52056 Aachen, Germany

Table of Contents

Topics in Current Chemistry, Vol. 184
© Springer Verlag Berlin Heidelberg 1996

Catalytic aldol reactions are among the most useful synthetic methods with a high potential for
convergent asymmetric synthesis. As an increasing number of lyases is becoming available for
preparative applications, in this review the state of the art concerning the application of these
biocatalysts for enzymatic C–C bond formations is evaluated. The scope, and the individual
limitations, of the most important types of enzyme-catalyzed transformations are discussed against
a more recent mechanistic and protein-structural background. Special emphasis is placed on the
synthetic power of a stereodivergent building-block approach which is facilitated by the prevalence
of families of stereocomplementary enzymes which often have a very similar, broad substrate
tolerance. The methodology is highlighted by exemplary applications to the synthesis of valuable
polyfunctionalized products which otherwise are difficult to prepare and to handle by classical
chemical methods.

1 Introduction

The importance of being able to synthesize enantiometrically pure compounds
(EPC syntheses) has continued to increase ever since Louis Pasteur about 150
years ago realized that molecular asymmetry causes optical activity not the least
through the — often grievous — experience that the biological activity, of enantiomers can differ dramatically in its kind and intensity because of the chiral
nature of life processes [1–3]. Current regulatory requests for an evaluation of

the potential benefit — or threat — of all enantio- and/or diastereoisomers of new xenobiotic agents, especially of pharmaceuticals, calls for economical means for their independent, directed synthesis.

As an alternative to classical chemical methodology, enzymes are finding increasing acceptance in modern chemical research and production as catalysts for the in-vitro synthesis of asymmetric compounds because they are intrinsically chiral and optimized by evolution for a high catalytic efficiency [4–23]. In contrast to most classical chemical techniques, biocatalytic conversions can usually be performed on underivatized substrates, thus making tedious and costly protecting-group manipulations superfluous [24], because of the high levels of selectivity offered by enzymes and because of the usually very mild reaction conditions that are compatible with most functional groups. In parallel with regulatory demands for pharmaceuticals and with increasing environmental concerns, biocatalysis offers an attractive option for the development of new economical and ecologically acceptable processes for the synthesis of compounds with high optical purity. Considerable advantages are thus evident for biologically-relevant classes of compounds — typically polyfunctional and water soluble – such as amino acids [10, 15] or carbohydrates [21, 25]. Partly for this reason, partly because of the medicinally relevant implications of oligomeric derivatives of these compounds in central biological recognition phenomena — such as cell-cell communication for cell adhesion, viral infection, or cell differentiation in organ development and carcinogenesis [26–29] — the chemistry of amino acids, peptides, and carbohydrates was the first to experience a profound impact of enzymatic catalysis. Pertinent to this flourishing field are advances in biotechnology such as techniques for the immobilization and stabilization of biocatalysts [30–33] or in-situ cofactor regeneration systems [34, 35] for economical application of co enzyme dependent enzymes, as well as major progress in recombinant DNA technology, which permit the inexpensive production of large quantities of enzymes [36] and the possible rational tailoring of their properties [37].

An asymmetric C–C coupling, one of the most important and challenging problems in synthetic organic chemistry, seems to be most appropriate for the creation of a complete set of diastereomers because of the applicability of a convergent, "combinatorial" strategy [38–40]. In Nature, such reactions are facilitated by lyases which catalyze the (usually reversible) addition of carbonucleophiles to C=O double bonds, in a manner mechanistically most often categorized as aldol and Claisen additions or acyloin reactions [41]. The most frequent reaction type is the aldol reaction, and some 30 lyases of the aldol type ("aldolases") have been identified so far [42], of which the majority are involved in carbohydrate, amino acid, or hydroxy acid metabolism. This review will focus on the current state of development of this type of enzyme and will outline the scope and limitations for their preparative application in asymmetric synthesis.

2 General Aspects

2.1 Classification of Lyases

Apart from their metabolic connection and possible mechanistic distinction, a number of these C–C forming catalysts can be grouped into certain families depending on the usage of a common nucleophilic component. For synthetic purposes, the three most useful families are the pyruvate-dependent lyases, the phosphoenolpyruvate-dependent lyases (PEP synthases), and the dihydroxy-acetone-phosphate-dependent lyases (DHAP lyases) which all add a three-carbon ketone fragment onto a carbonyl group of an aldehyde (Fig. 1). Members of the first two classes produce 3-deoxy-2-oxoacids and thereby generate a single stereocenter, while members of the latter family form 3,4-dihydroxyketone 1-phosphates which contain two new vicinal chiral centers at both termini of the created C–C bond.

Mechanistically, the activation of the aldol donor substrates by stereo-specific deprotonation is achieved in two different ways (Fig. 1) [43]. Class I

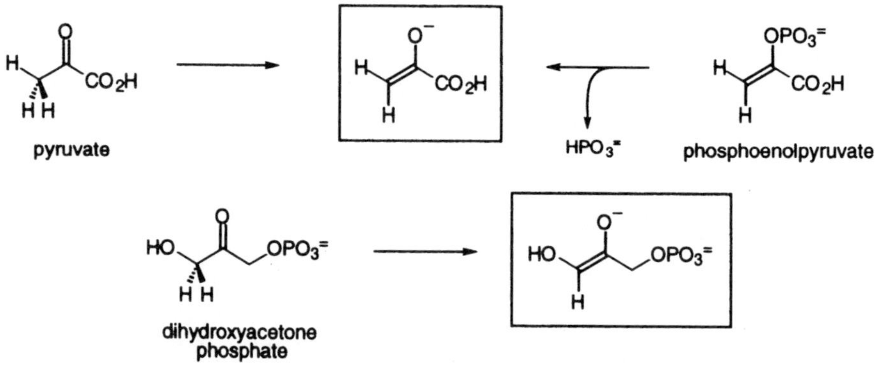

Scheme 1. Classification of lyases according to their donor components

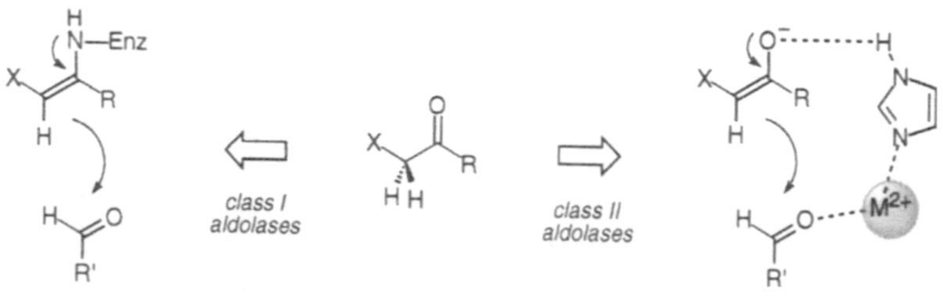

Fig. 1. Mechanistic distinction of aldolases according to Schiff-base (class I) or metal-ion (class II) activation of substrates

aldolases bind their substrates covalently via imine/enamine formation to an active site lysine residue for initiating bond cleavage or formation (Fig. 2). This can be demonstrated by reductive interception of the intermediate with borohydride which causes irreversible inactivation of the enzyme due to alkylation of the amine [44–46]. In contrast, class II aldolases utilize transition metal ions as a Lewis acid cofactor which facilitates deprotonation by a bidentate coordination of the donor to give the enediolate nucleophile (Fig. 3). An earlier model in which it was proposed that the aldehyde coordinates directly to the metal while the donor enolate is protonated by histidine (Fig. 1) [47] no longer

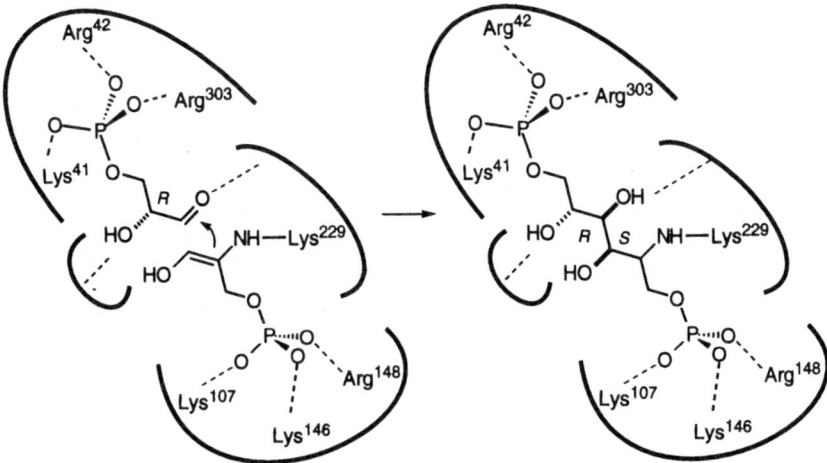

Fig. 2. Schematic representation of substrate binding and C–C bond formation for the class I fructose 1,6-bisphosphate aldolase from rabbit muscle

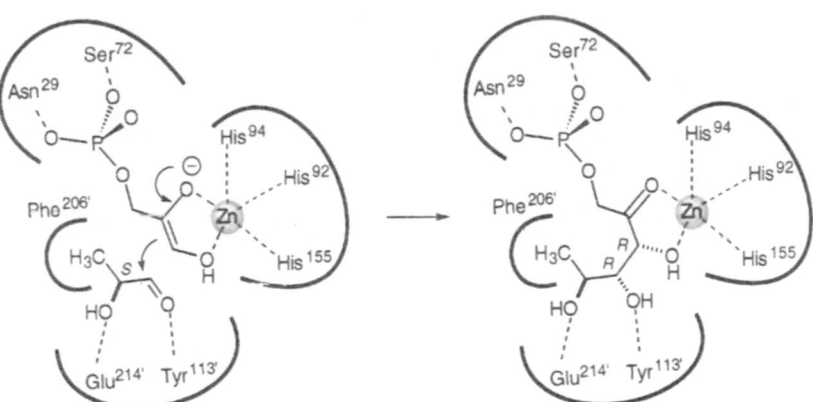

Fig. 3. Schematic representation of substrate binding and C–C bond formation for the class II fuculose 1-phosphate aldolase from *Escherichia coli*

holds in the light of a very recent protein crystal structure determination (cf. Sect. 5.5). Mostly, this effect is achieved by a tightly bound Zn^{2+} ion but some other divalent cations may act in its place. Evidently, aldolases of the latter class can be effectively inactivated by the addition of strong complexing agents such as EDTA [43].

In addition to the larger families of preparatively useful aldolases, some less common aldolases have been evaluated lately for preparative use. A range of mechanistically distinct enzymes, which are actually categorized as transferases but which also catalyze aldol-related additions through the aid of cofactors such as pyridoxal 5-phosphate (PLP), thiamine pyrophosphate (TPP), tetrahydrofolate (THF), or coenzyme A (CoA), are becoming more frequently applied in organic synthesis. Because of their emerging importance and/or commercial availability, a selection of these enzymes and examples of their synthetic utility will be included in further separate chapters.

The biochemical nomenclature of enzymes [48] follows reaction classifications that are different from the needs of synthetic organic chemists. While the common EC numbers are valuable for reference, often neither the clumsy correct terminology nor the often employed shorthand notations derived from old (and now often incorrect) trivial substrate names are very helpful. For many useful biocatalysts this has led to a situation where many different acronyms have been utilized in parallel by different authors for the same enzyme, or where the same acronym has been used for enzymes from different organisms that share a common natural reference reaction but actually differ considerably in their catalytic capability, which may lead to additional confusion for the non-specialist. In an attempt to clarify this situation, we propose a new scheme for structuring enzyme designation in a simplified fashion that draws from the nomenclature used by geneticists for structural genes [49]. The latter system utilizes an italicized three-letter code indicating the biological function of the gene product, usually a reference to the metabolic system that the corresponding protein is involved in, followed by a single capital letter for additional indexing. In a similar manner, enzymes may be conveniently termed according to the reference substrates (which often falls into a compound category for which an IUPAC-recommended three-letter abbreviation already exists, such as standard amino acids or carbohydrates), where functional groups and the requirements of relative stereochemistry would be immediately visible (often this will coincide with genetic designations for the underlying genes). This would be followed by a single capital to indicate the type of conversion that the enzyme catalyzes and which may be derived from the current designation of enzyme classes. Among the synthetically most useful classes of biocatalysts are certainly dehydrogenases (D), oxidases (O), kinases (K), (glucosyl)transferases (T), aldolases (A), synthases (S), isomerases (I), or epimerases (E). In order to further differentiate additional substrate specificities, if needed, stereochemical descriptors (R/S or D/L) or numerical indexes can precede or be included, as is already common usage for the regioselectivity observed with glycosyltransferases towards an acceptor substrate. Species origin may be indicated by a subscripted index, which for microbial origins should follow the guidelines in practice, e.g. for restriction enzymes [50]. To clearly distinguish from genetic usage, enzyme designations would have to be in normal type.

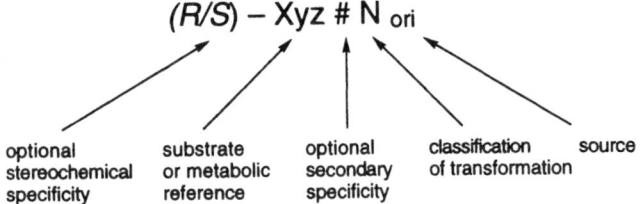

For example, some well-known enzymes utilizing glucose as a substrate would be designated as GlcD for glucose dehydrogenase, GlcO for glucose oxidase, GlcK for glucokinase, GlcT for glucosyltransferase, or GlcI for glucose isomerase. Differentiation of fucosyltransferases would be given as Fuc1,3T versus Fuc1,4T to indicate a specificity for the 3-O or 4-O function of the acceptor pyranose, and a distinction of the tagatose 1,6-bisphosphate aldolases from *Escherichia coli* or *Staphylococcus aureus* would be evident from TagA_{eco} and TagA_{sau} descriptions, respectively. We realize that this system may have its drawbacks, particularly in naming enzymes for which no recommended substrate code is available or in those cases where potential ambiguities result from very similar codes such as those for glucose (Glc) and glutamic acid (Glu) that may be less familiar to organic chemists, or by a possible necessity for them to become familiar with microbiological terminology. However, it may be envisaged that a more general and systematic shorthand description of enzymes according to their function — for example the fructose 1,6-bisphosphate aldolase from rabbit muscle would be termed FruA_{rab} instead of less precise 'aldolase', FDP aldolase, FBP aldolase, Fru aldolase, or RAMA that are used synonymously in the current literature — will help in making the field more transparent for novices and specialists alike. For several reasons it has been impossible to switch completely from the previous enzyme notations in this review but, on an experimental basis, we have tried to cover the most important cases for illustration.

2.2 Enzyme Selectivities

Typically, because of mechanistic requirements the lyases are highly specific for the nucleophilic donor component. This includes the necessity for a reasonably high substrate affinity as well as the general difficulty of binding and anchoring a rather small molecule in a fashion that restricts solvent access to the carbanionic site after deprotonation and shields one enantiotopical face of the nucleophile in order to secure correct diastereofacial discrimination (Figs. 2 and 3) [51]. Usually, approach of the aldol acceptor to the enzyme-bound nucleophile occurs stereospecifically following an overall retention mechanism, while the facial differentiation of the aldehyde carbonyl is responsible for the relative stereoselectivity. In this manner, the stereochemistry of the C–C bond formation is completely controlled by the enzyme, often irrespective of the constitution or configuration of the substrate. Only few specific exceptions to this rule are known at present and these will be discussed individually below.

On the other hand, most of the lyases allow a reasonably broad variation of the electrophilic acceptor component, which usually is an aldehyde. This feature, which nicely complements the emerging trend of combinatorial synthesis [38–40], makes possible a stereodivergent strategy for the synthesis of groups of stereoisomeric compounds by employing separate enzymatic catalysts to selectively produce individual diastereomers at will from the same starting material (Sect. 7).

Although a considerable amount of research has been done using the available techniques of enzymology, site-directed mutagenesis, and protein crystallography, for none of the aldolases was it possible to devise a conclusive model that would account for all the requisite roles of known protein residues in the individual steps of catalysis; however progress is being made [52–54]. Hopefully, an emerging understanding of molecular recognition in aldolase–substrate interactions during a catalytic cycle will enable these catalysts to be engineered with the aim of improving substrate tolerance and stereoselectivity for asymmetric syntheses, or in order, in the more distant future, to rationally redesign lyases for novel catalytic functions.

2.3 Technical Considerations

Although the majority of those aldolases that are attractive for synthetic applications stem from catabolic pathways where they function in the degradative cleavage of metabolites, the reverse C–C bond-forming processes are often favored by thermodynamic relations. Because of the bimolecular nature of the reaction, the product fraction at equilibrium may be increased in less favorable cases by working at higher substrate concentrations or by driving the reaction with a higher concentration of one of the reactants. Individual choice will certainly depend on the cost of starting materials, but factors such as enzyme inhibition by substrate(s) or product may be critical as well. The latter factor will be more obvious if one recognizes that for most of the lyases both the donor and acceptor components contain strong electrophilic sites as aldehyde or ketone carbonyl groups, and that many lyases, including class I aldolases, involve a covalent binding of substrate and product at — but not necessarily restricted to — the active site.

The majority of useful lyase families utilize anionically functionalized substrates such as pyruvate or dihydroxyacetone phosphate which remain unaltered during catalysis. The charged group thereby introduced into the products (phosphate, carboxylate) not only constitutes a handle for binding of the substrates by the enzymes but also can facilitate the preparative isolation from aqueous solution of the products and their purification by salt precipitation or ion exchange techniques. One problem arising from the affinity of the enzymes to anionic substrates is the potential (competitive) inhibition by common buffer salts, e.g. by inorganic phosphate.

For routine practical application, most of the aldolases are sufficiently robust to allow them to be used in solution for an extended period of time, often for several days. To enhance the lifetime and to facilitate recovery of the biocatalysts after completion of the desired conversion, several options have been tested to immobilize the enzymes to or within insoluble matrices [55, 56], including cross-linking of enzyme crystals [57, 58], or to confine them in membrane reactors [56, 59–61]. Applications of current techniques are indicated with the individual enzymes, if available. For synthetic simplification, techniques of applying recombinant whole cells, which overexpress the appropriate lyase in abundance, seem attractive at first glance because cell metabolism could in principle be utilized for providing the aldol donor in situ for the desired addition reaction. However, the central implication of these donor components in several general metabolic pathways, including that of sustaining cellular viability, is a major disadvantage, particularly for those synthetic systems where product formation is not strongly favored by thermodynamic relationships.

3 Pyruvate Lyases

In vivo, pyruvate lyases perform a catabolic function. The synthetically most interesting types are those involved in the degradation of sialic acids or the structurally related octulosonic acid KDO, which are higher sugars typically found in mammalian or bacterial glycoconjugates [62–64], respectively. Also, hexose or pentose catabolism may proceed via pyruvate cleavage from intermediate 2-keto-3-deoxy derivatives which result from dehydration of the corresponding aldonic acids. Since these aldol additions are freely reversible, the often unfavourable equilibrium constants require that reactions in the direction of synthesis have to be driven by an excess of one of the components, preferably pyruvate for economic reasons, in order to achieve a satisfactory conversion.

3.1 N-Acetylneuraminic Acid Aldolase

N-Acetylneuraminic acid aldolase (Neu5NAc aldolase or NeuA; EC 4.1.3.3), also known as sialic acid aldolase, catalyzes the reversible addition of pyruvate

(2) to N-acetyl-D-mannosamine (ManNAc, 1) to form the parent sialic acid N-acetylneuraminic acid (Neu5NAc, 3). The NeuA lyases found in both bacteria and animals are type I enzymes that form an enamine intermediate with pyruvate to promote a *si*-face attack to the carbonyl group of the aldehyde acceptor with formation of a new stereogenic center of absolute (4S) configuration [65, 66]. Sugar substrates 1 and 3 in cleavage and synthesis directions are specifically bound in their respective α-anomeric forms [67, 68].

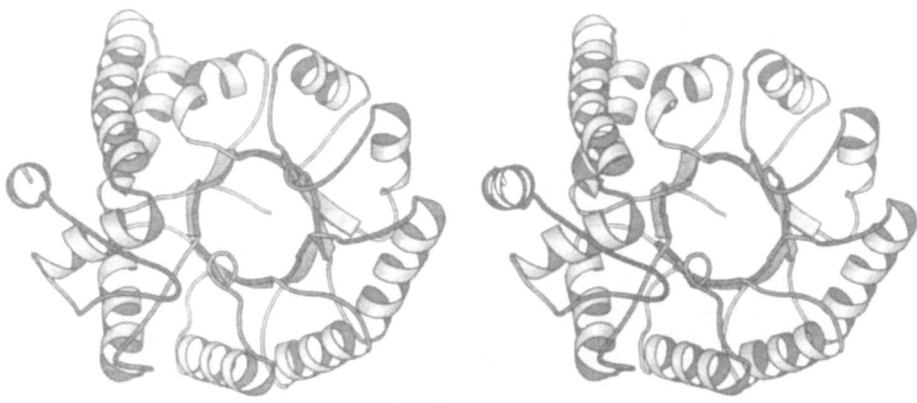

Enzyme preparations from *Clostridium perfringens* and *Escherichia coli* are commercially available, and the latter enzyme [69] has been cloned, overexpressed [70–72], and crystallized [73]. Its spatial structure has been determined recently by X-ray crystallography at 2.2 Å resolution which has revealed that the enzyme is a tetramer and belongs to the α/β-barrel class of proteins [74] (Fig. 4). The active site pocket is located at the carboxy-terminal end of the eight-stranded β-barrel, and the reactive lysine residue, forming a Schiff base with pyruvate, has been identified to be Lys165 (Fig. 5). However, the current X-ray model excludes the catalytic participation of a histidine residue [74],

Fig. 4. Stereo ribbon plot [76] of a NeuA subunit viewed down the β-barrel axis from its carboxy-terminal end

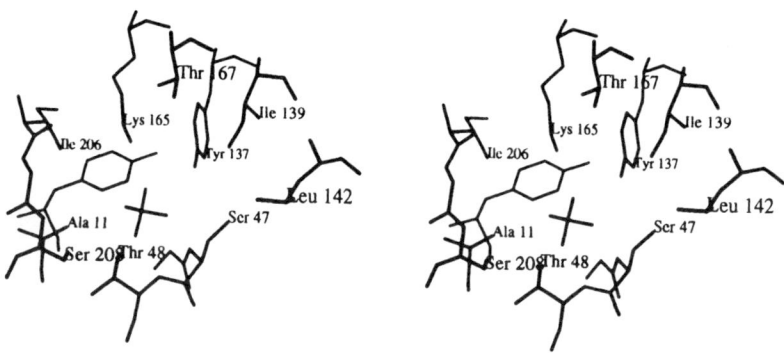

Fig. 5. Stereo view of the putative active site of NeuA showing the environment of the Schiff-base-forming Lys165 residue

which earlier had been proposed to be involved in the substrate binding and polarization of the acceptor carbonyl [68, 75]; rather, it is likely that this role is adopted by a tyrosine residue (Tyr137; cf. mechanistic views for other aldolases). Additional electron density is presumed to be a sulphate ion which occupies the carboxylate binding site.

Because of its earlier availability, the *Clostridium* specimen in particular has been extensively investigated for synthetic purposes both in its soluble and immobilized forms. The enzyme has a broad pH optimum around 7.5 and useful stability in solution at ambient temperatures [77]. Because of its biological importance in cellular recognition events [64] mediated by mammalian glycoconjugates such as cell adhesion [27, 78–80] or viral infections [81], a practical route to the synthesis of natural neuraminic acid **3** or derivatives and, particularly, that of its unnatural analogs is a crucial target of synthetic and medicinal chemistry. Preparation of **3** has been successfully demonstrated by several groups on the multi-gram scale [60, 77, 82, 83]. As the equilibrium constant of 28.7 M^{-1} at 25°C in the direction of synthesis does not strongly support product formation [60], usually a 7–10-fold excess of pyruvate has been applied to achieve a preparatively useful conversion of the more valuable sugar derivative **1**. Product isolation is complicated by the requirements of ion exchange chromatography in order to achieve separation from excessive pyruvate equivalents. To facilitate work-up of reaction mixtures, pyruvate decarboxylase present in yeast cells has been applied to the selective destruction of residual pyruvate into the volatiles acetaldehyde and CO_2 (cf. Scheme 22) [84]. An even further reduction — or even elimination — of pyruvate excess would be possible by coupling the synthesis of **3** to a thermodynamically favored process; indeed, this has recently been realized by combining the formation of **3** with a multienzyme system designed for a sialyltransferase-catalyzed synthesis of a sialosaccharide [85], where the driving force is provided by consumption of **3** in the CTP-dependent formation of the glycosyl donor

CMP-Neu5NAc and by irreversible in situ regeneration of the nucleoside triphosphate.

The expensive substrate ManNAc **1** can be produced by a separate base-catalyzed epimerization of *N*-acetyl-D-glucosamine (GlcNAc, **4**) [83]. A more elaborate approach has been pursued for scale-up using an enzyme membrane reactor (EMR) where in a combined enzymatic protocol the isomerization of **4** was integrated by the usage of a commercial, ATP-activated *N*-acyl-glucosamine 2-epimerase (EC 5.1.3.8) [60]. Both enzymes were confined to an EMR which was then continually charged with a solution containing **4** and pyruvate at a reduced 1 : 1.9 ratio to produce neuraminic acid **3** with a high space-time yield without enzyme deactivation. Most recent scale-up developments have yielded an industrial process by which neuraminic acid can be readily made in ton quantities using the recombinant NeuA from *E. coli* in immobilized form [86].

The specificity of the aldolase for pyruvate seems to be absolute since all of the structural variants tested so far proved inactive, including 3-hydroxy or 3-halo derivatives or an analogous phosphonate [77]. On the other hand, the enzyme displays a fairly broad tolerance for analogs of the electrophilic substrate such as a number of sugars and their derivatives larger or equal to pentoses, but small open-chain aldehydes such as glyceraldehyde are not acceptable. A number of natural and unnatural sialic acid derivatives have been prepared by replacement of the natural D-*manno*-configurated substrate with aldose derivatives containing modifications such as epimerization, substitution, or deletion at positions C-2, -4, or -6 (Table 1). Most notably, the *N*-acetyl group in **1** may either be omitted [87, 88] or be replaced by sterically-demanding substituents such as a branched-chain homologue [89], *N*-Cbz [90, 91] or even nonpolar phenyl groups [88] without destroying activity; epimerization at C-2, however, is restricted to small polar substituents such as OH (e.g. D-glucose) or fluoride at strongly decreased reaction rates. Large acyl substituents are also tolerated at C-6 as shown by the conversion of a BOC-glycyl derivative **5** as a precursor to a fluorescent sialic acid conjugate [92] of a type that may facilitate investigations of sialyltransfer [93–95] or adhesion [96]. 9-Acyl-substituted sialic acids such as **6** occur naturally as tissue and development specific variants of neuraminic acid [97].

In most cases, a high level of asymmetric induction for the (4*S*) configuration is retained. However, a number of carbohydrates were also found to be converted with random stereoselectivity for the generation of the C-4 configuration (Fig. 6) [59, 84, 104]. In certain cases, ultimately a complete inversion of the normal stereopreference and exclusive formation of (4*R*)-configurated adducts can occur [105]. Clearly, product composition depends on conversion, which indicates a contribution of thermodynamic control. This outcome has been interpreted to result from substrate control of stereoselectivity following an anti-Cram rather than Cram-type mode of aldol addition [84]. However, consideration of the Cram transition state model is totally inapplicable to reactions in which in fact a chiral enzymic catalyst controls relative kinetics of product formation. Since substrates are initially bound by the enzyme in their cyclic forms before ring opening and aldolization occurs, thus an alternative interpretation seems more credible, i.e. one that considers the inverse conformational preferences of the different substrate classes [59]. Whereas **1** and related substrate analogs that give only normal products exclusively adopt a $^4C_1$ chair conformation (e.g. D-mannose), substrate analogs leading to stereoerror formation seem to prefer predominantly the inverted $^1C_4$ conformers (e.g. L-mannose).

Table 1. Substrate tolerance of neuraminic acid aldolase

R₁	R₂	R₃	R₄	R₅	Yield [%]	Rel. rate [%]	Ref.
NHAc	H	OH	H	CH₂OH	85	100	[60, 77, 82, 83]
NHAc	H	OH	H	CH₂OAc	84	20	[77, 82]
NHAc	H	OH	H	CH₂OMe	59	–	[82]
NHAc	H	OH	H	CH₂N₃	84	60	[87]
NHAc	H	OH	H	CH₂OP(O)Me₂	42	–	[87]
NHAc	H	OH	H	CH₂O(L-lactoyl)	53	–	[82, 87]
NHAc	H	OH	H	CH₂O(Gly-N-Boc)	47	–	[92]
NHAc	H	OH	H	CH₂F	22	60	[87]
NHAc	H	OMe	H	CH₂OH	70	–	[82, 98]
NHAc	H	N₃	H	CH₂OH	46	–	[99]
NHAc	H	H	H	CH₂OH	70	–	[88, 98]
NHC(O)CH₂OH	H	OH	H	CH₂OH	61	–	[82]
NHCbz	H	OH	H	CH₂OH	75	–	[90, 91]
NHC(S)Me	H	OH	H	CH₂OH	56	–	[100]
N₃	H	OH	H	CH₂OH	80	13	[90, 101–103]
CH₂NHAc	H	OH	H	CH₂OH	75	–	[89]
OH	H	OH	H	CH₂OH	84	91	[84, 88]
OH	H	H	H	CH₂OH	67	35	[88]
OH	H	H	F	CH₂F	40	–	[87]
OH	H	OH	H	H	66	10	[88]
H	OH	OH	H	CH₂OH	28	7	[103]
H	F	OH	H	CH₂OH	30	–	[87]
H	H	OH	H	CH₂OH	36	130	[87, 88]
Ph	H	OH	H	CH₂OH	76	–	[88]

A critical and distinctive factor seems to be recognition of the configuration by the enzymic catalyst at C-3 in the aldehydic substrate [59].

Thus, a generalized mechanistic rationale for NeuA catalysis can be advanced as outlined in Scheme 2, which, for aldopentose or aldohexose acceptor substrates in a pyranoid form (**B**), considers the three primary substrate recognition sites demanded by the α-anomeric specificity, distinction of the 3-hydroxyl function, and proton donation to the intracyclic oxygen atom. The latter is necessary to assist in the ring opening of the sugar which has to be catalyzed at the active site prior to nucleophilic attack by the pyruvate enolate; deprotonation of the anomeric hydroxyl group by an active site base will assist in the ring opening as well as, via the resultant hydrogen bonding of the open-chain aldehyde intermediate, in polarizing the aldehyde carbonyl for enhanced electrophilicity. Accordingly, the (3S)-α-⁴C₁ structure of "normal" substrates (**B**) determines the unbiased pathway for si-face attack, leading to products of the sialic acid type, while the corresponding β-anomer has only low binding affinity (**A**) and remains intact. Conversely, stereoregular si-attack demands that sub-

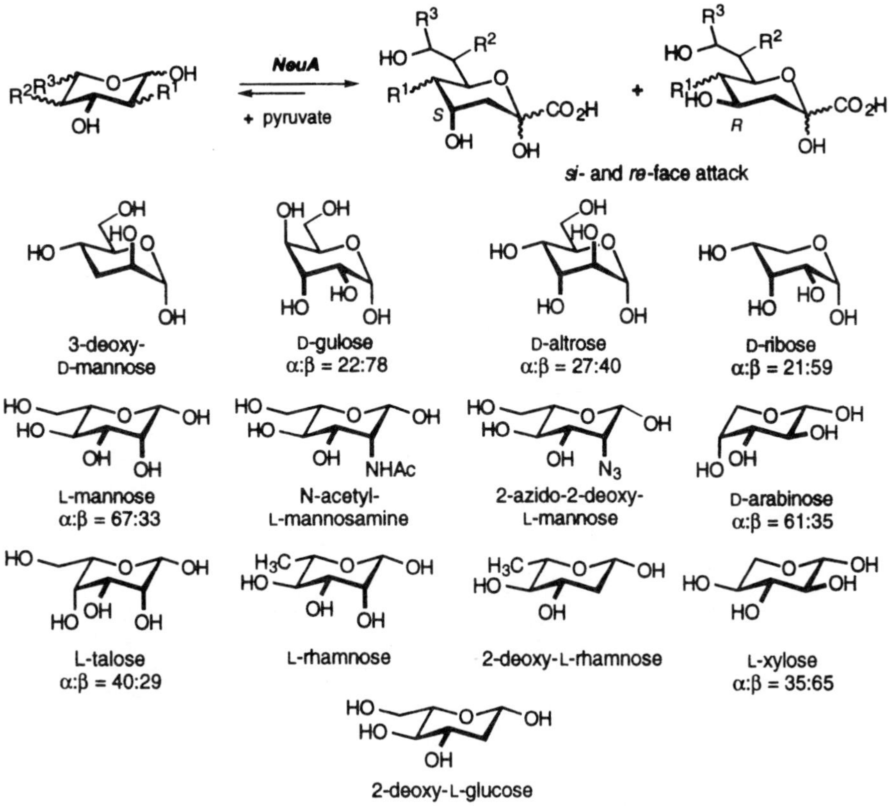

Fig. 6. Limited stereoselectivity in NeuA-catalyzed aldol additions, observed with aldopyranose substrates (including known anomeric ratios)

strates inversely configured at C-3 must react from their $(3R)$-α-$^1C_4$ structure, which is less well accommodated in the active site (**C**). A comparison of kinetic parameters for various carbohydrate substrates indeed shows that substrates with normal stereoselectivity react faster than corresponding ones with lack of or with inverted stereoselectivity [59]. An inverted $(3R)$-α-$^4C_1$ chair conformation is strongly disfavored by 1,3-diaxial oxygen repulsion and would also lack binding affinity (**F**). After anomerization to the β-form, $(3R)$-configured compounds will likely be bound and converted from the $(3R)$-β-$^4C_1$ conformation along the re-face trajectory caused by a reverted positioning of the aldehyde moiety relative to the catalytic environment (**E**). This crude model also suggests that the $(3R)$-β-$^1C_4$ conformer is probably inactive (**D**). However, the differing prevalence of anomers in solution (Fig. 6) and the individual contributions of additional substituents of minor importance to binding are expected to modify the relative transition state energies and thus the kinetics along competing reaction channels. Experimentally, relative kinetics for re- versus si-product formation indeed can vary considerably [59, 84], whereas final product ratios

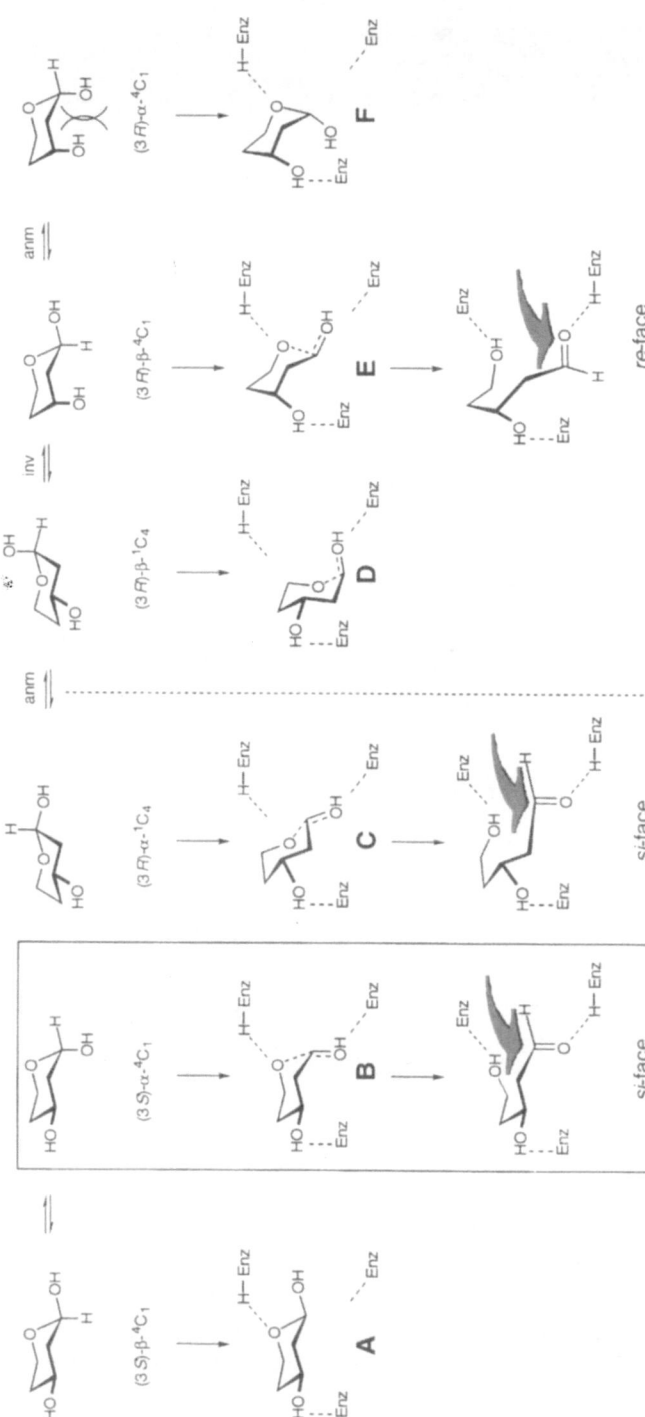

Scheme 2. Mechanistic analysis of reaction pathways for NeuA catalysis that considers the crucial influences of C-3 and anomeric configurations and of chair conformations on the three-point attachment of aldoses as acceptor substrates. *Si*-face attack leads to regular (4S)-configurated adducts while *re*-face attack leads to inverted stereochemistry (abbreviations: anm = anomerization, inv = ring inversion)

are largely determined by relative thermodynamic product stabilities due to the reversibility of the aldolization. For specific cases, however, literature reports of long-term product ratios do not reflect true thermodynamic relationships. For example, from the reaction of L-mannose (Scheme 3) leading to adducts 7 and 8, the early ratio of 64:36 (1.5 h) slowly converts to 93:7 (7 d) [84] with the regularly configurated 7 ultimately vanishing completely [105]. This may reflect the fact that for the reverse reaction some inversely configurated products either may not be able to bind to the enzyme productively, or that the catalyzed back reaction in certain cases would have to proceed via very unfavorable transition states, such as sterically crowded or boat conformations (cf. Sect. 5.2), thus causing apparent irreversibility.

Products generated by the unusual *re*-face attack for (4R)-stereochemical preference include a number of related higher ulosonic acids of biological importance such as D-KDO prepared from D-arabinose [59, 84, 88], or the formation of the enantiomers of naturally occurring sugars, such as L-Neu5NAc, L-KDN, and L-KDO from L-*N*-acetylmannosamine, L-mannose, and L-arabinose [84], respectively. Ready access to compounds of this type may be particularly valuable for investigations into the biological activity of sialoconjugates containing nonnatural sialic acid derivatives [106].

The deviating behaviour of biological catalysts from predictable patterns stresses the fact that the development of enzymatic processes is largely an empirical endeavor. It must be emphasized that the observed but less desirable

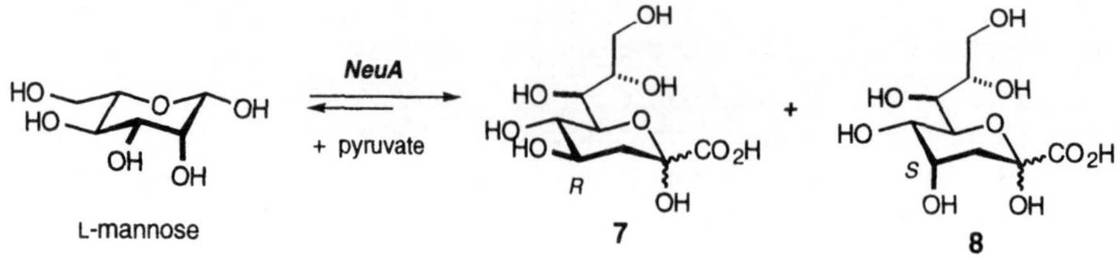

7	:	8	time
100	:	57	1.5 h
100	:	7	7 d
100	:	0	∞

Scheme 3. Time-dependent inversion of stereochemistry for NeuA-catalyzed aldol addition with L-mannose during long-time equilibration

reduction — or even inversion — of stereoselectivity of the NeuA is by no means unique (cf. Sect. 5) and, in fact, is becoming more and more likely whenever an enzyme, optimized by Nature for the binding of a specific substrate, is confronted with substrate analogs that are structurally grossly different from their natural counterparts.

3.2 2-Keto-3-deoxyoctosonic Acid Aldolase

The functionally-related enzyme 2-keto-3-deoxyoctosonate aldolase (KDO aldolase or KdoA; EC 4.1.2.23), correctly termed 3-deoxy-D-*manno*-octulosonic acid aldolase, is an inducible enzyme found among a variety of Gram-negative microorganisms [107] where D-KDO **9** is a core constituent of the outer membrane lipopolysaccharide [62, 63]. The aldolase, which reversibly degrades **9** to D-arabinose **10** and pyruvate, had originally been isolated from *Aerobacter cloacae* and had been successfully used for a small-scale synthesis of specifically labeled **9** from **10** and labeled pyruvate [108]. Preparative applications suffer from an unfavorable equilibrium constant of 13 M^{-1} in direction of synthesis [108]. A more recent screening for a convenient enzyme source has encountered severe difficulties because the enzyme seemed to be localized in the cell wall or membrane fraction [107]. Lately, the *Aureobacterium barkerei* strain KDO-37-2 has been shown to contain enhanced levels of KDO aldolase when grown on KDO medium (~25 U from 2 L culture) and partially purified enzyme preparations have been studied for synthetic applications [109]. Similar to the NeuA, the KdoA enzyme has a broad substrate specificity for aldoses (Table 2) while pyruvate was found to be irreplaceable. As a noteable distinction, KdoA was also active on smaller acceptors such as glyceraldehyde. Once available in quantity, the enzyme may be useful in the synthesis and development of new KDO analogs with potential antibacterial activity [110, 111].

The stereochemical course of aldol additions generally seems to adhere to a *re*-face attack on the aldehyde carbonyl, a facial selection complementary to that of sialic acid aldolase. On the basis of the results published so far it may further be concluded that a $(3R)$-configuration is necessary (but not sufficient), and that stereochemical requirements for C-2 are less stringent.

Table 2. Substrate tolerance of 2-keto-3-deoxyoctosonic acid aldolase [109]

Substrate	R_1	R_2	R_3	Yield [%]	Rel. rate [%]
D-altrose	OH	H	(RR)-$(CHOH)_2$–CH_2OH	–	25
L-mannose	H	OH	(SS)-$(CHOH)_2$–CH_2OH	61	15
D-arabinose	OH	H	(R)-CHOH–CH_2OH	67	100
D-ribose	H	OH	(R)-CHOH–CH_2OH	57	72
2-deoxy-2-fluoro-D-ribose	F	H	(R)-CHOH–CH_2OH	19	46
2-deoxy-D-ribose	H	H	(R)-CHOH–CH_2OH	47	71
5-azido-2,5-dideoxy-D-ribose	H	H	(R)-CHOH–CH_2N_3	–	15
D-threose	OH	H	CH_2OH	–	128
D-erythrose	H	OH	CH_2OH	39	93
D-glyceraldehyde	H	OH	H	11	23
L-glyceraldehyde	OH	H	H	–	36

3.3 2-Keto-3-deoxy-6-phospho-D-gluconate Aldolase ≡ 2-Keto-4-hydroxyglutarate Aldolase

The 2-keto-3-deoxy-6-phospho-D-gluconate (**11**) aldolase (KDPGlc aldolase or KdgA; EC 4.1.2.14) provides the basis of the Entner-Doudoroff pathway that is used by many species of bacteria for the degradation of 6-phosphogluconate into pyruvate and D-glyceraldehyde 3-phosphate **12**. The equilibrium constant is in favor of synthesis (10^3 M^{-1}) [112]. Enzymes from disparate microbial sources can be readily isolated by two-step differential dye-ligand chromatography [113, 114]. The substrate **11** can be obtained from gluconic acid by microbial synthesis employing an *Alcaligenes eutrophus* mutant lacking KdgA activity [115]. The enzymes from *E. coli* [116–118], *Zymomonas mobilis* [119], and *Erwinia chrysanthemi* [120] have recently been cloned. The complete protein sequence has been reported [121] for the KdgA enzyme isolated [122] from *Pseudomonas putida* and its X-ray crystal structure has been solved at 2.8 Å resolution, revealing that the class I enzyme is a trimer of subunits which fold to form an eight-stranded α/β-barrel structure [123]. The putative azomethine-forming lysine residue Lys144, identified earlier by chemical derivatization [121], is located on a loop at the carboxyl end of the β-barrel.

The catabolism of hydroxyproline proceeds through 4-hydroxy-2-oxo-glutaric acid **13** that is retroaldolized to pyruvate **2** and glyoxalate **14** by an enzymatic activity termed 2-keto-4-hydroxyglutarate aldolase (KHG aldolase or KhgA; EC 4.1.3.16) which is found in microbial and mammalian organisms [124–129]. The amino acid sequence, including the active site lysine residue for Schiff base formation, has been determined for the *E. coli* enzyme [130] which more recently has also been cloned and overexpressed [131].

However, it came as a great surprise when it was discovered that the DNA sequences of both KdgA and KhgA of *E. coli* proved to be identical and that in fact both activities for separate metabolisms reside within the same protein [116]. The most recent proof came from analysis of the enzyme from *Azotobacter vinelandii* which was shown to be bifunctional for cleavage of **11** as well as of **13** with the substrates sharing the same active site [132]. In both cases, the enzyme has to recognize identical (4S)-configurations; in fact, stereochemical investigations with specifically labeled [3-³H,²H,H]-pyruvate suggest that configuration at C-3 of pyruvate is retained during C–C synthesis, i.e. both the exchanging proton and aldehyde approach the same face of bound enolpyruvate [133, 134].

Controversy remains in the determination of substrate tolerance for KdgA/KhgA aldolases from different sources. Early assay studies with KhgA prepared [127–129] from rat liver concluded that the catalyst had an unusually wide ranging tolerance for nucleophilic components, including a number of 3-substituted pyruvate derivatives as well as pyruvaldehyde, acetaldehyde, and pyruvic esters [135]. Later, other workers using enzymes from rat or bovine liver and from *E. coli* reported their inability to reproduce these results but noted a rather limiting specificity [136].

The *Pseudomonas fluorescens* KdgA was shown to accept several polar-substituted aldehydes, albeit at rates much lower (< 1%) than the phosphorylated natural substrate 12 (Table 3) [137]. Simple aliphatic or aromatic aldehydes were not converted. Synthetic utility and high stereoselectivity with unnatural substrates were demonstrated by conversion of both the D-configurated glyceraldehyde (D-15) and lactaldehyde (D-16) to form the respective (4S)-configurated adducts 17 and 18 at the mmol scale.

The *E. coli* derived KhgA [138] was applied to prepare both the natural substrate L-(S)-13 (> 95% ee) and its antipode D-(R)-13 (60% ee) selectively on a mmol scale (Scheme 4) [136]. The unfavorable equilibrium constant of

Table 3. Substrate tolerance of 2-keto-3-deoxy-6-phospho-D-gluconate aldolase [137]

Substrate	R	Rel. rate [%]
D-glyceraldehyde 3-phosphate	D-CH_2OH–$CH_2OPO_3^=$	100.
3-nitropropanal	CH_2–CH_2NO_2	1.6
chloroethanal	CH_2Cl	1.0
D-glyceraldehyde	D-CHOH–CH_2OH	0.8
D-lactaldehyde	D-CHOH–CH_3	0.2
D-ribose 5-phosphate	D-*ribo*-$(CHOH)_3$–$CH_2OPO_3^=$	0.04
erythrose	*erythro*-$(CHOH)_2$–CH_2OH	0.01
glycolaldehyde	CH_2OH	0.01

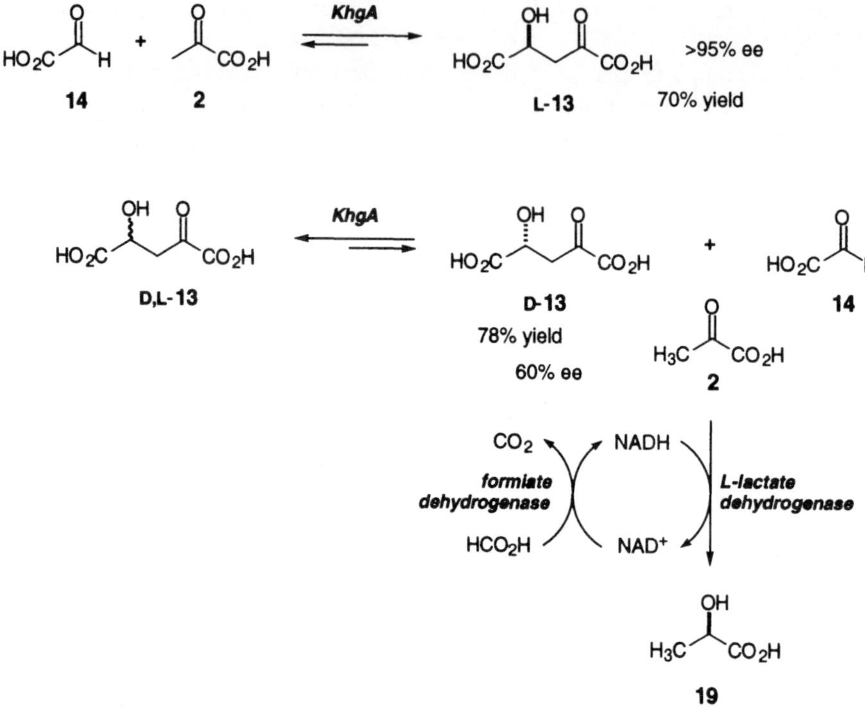

Scheme 4. Directed syntheses of both enantiomers of 4-hydroxy-2-oxoglutarate

0.76 mM^{-1} made it necessary to carry out the addition reaction at a high substrate concentration in order to achieve complete formation of the (4S)-configured L-enantiomer. The D-13 was obtained by selective decomposition of the L-enantiomer from a racemic mixture. In the latter case, in order to drive the equilibrium towards the cleavage products, pyruvate had to be removed by reduction to lactate (19) with L-lactate dehydrogenase (EC 1.1.1.27) supplemented by formate dehydrogenase (EC 1.2.1.2) for cofactor regeneration in situ.

3.4 Related Pyruvate Aldolases

Comparable to the situation for the sialic acid and KDO lyases (*vide supra*), sets of stereochemically complementary pyruvate lyases are known, e.g. in *Pseudomonas* strains, which act on related 2-keto-3-deoxy-aldonic acids [112]. The enzymes cleaving six-carbon sugar acid phosphates — the KdgA and 2-keto-3-deoxy-6-phospho-D-galactonate (20) aldolases (KDPGal aldolase; EC 4.1.2.21) [139] — are typified as class I enzymes, whereas those acting on non-phosphorylated five-carbon substrates — 2-keto-3-deoxy-L-arabonate (21) (KDAra aldolase; EC 4.1.2.18) [140, 141] and 2-keto-3-deoxy-D-xylonate (22) aldolase (KDXyl aldolase; EC 4.1.2.28) [142, 143] — are class II enzymes requiring a divalent metal cation for catalytic function, probably Mn^{2+}. All four types appear to be highly selective even for the acceptor components, that is, D-glyceraldehyde 3-phosphate **12** or glycolaldehyde **23**, and thus have been successfully applied only in stereoselective syntheses of their respective natural substrates. 2-Keto-3-deoxy-D-gluconate has been prepared by addition of pyruvate to glyceraldehyde using extracts or whole cells of the fungus *Aspergillus niger* containing a KDGlc aldolase (EC 4.1.2.n; possibly identical to KDXyl aldolase) [144, 145]. However, product mixtures also contained the *threo* diastereomer which either suggests a lack of enzyme stereoselectivity or points to the presence of a second aldolase having opposite facial selectivity. Indeed, such an enzyme (KDGal aldolase; EC 4.1.2.n; possibly identical to KDAra aldolase) was found in crude extracts of *Aspergillus terreus* grown on D-galactonate [146] and was used for the stereoselective preparation of 2-keto-3-deoxy-D-galactonate and related seven-carbon analogs. [147] An enzyme (EC 4.1.2.20) acting on 2-keto-3-deoxy-D-glucaric acid **24** in vivo with formation of pyruvate **2** and tartronic semialdehyde **25** has been found in microorganisms. The enzyme isolated from *E. coli* was described to show a potentially broad tolerance for C_2-C_4 substrate analogs [148].

4 Phosphoenolpyruvate Lyases ('Synthases')

Three types of lyases have been identified that catalyze the addition of phosphoenolpyruvate (PEP) to aldoses or to terminally phosphorylated sugar derivatives. With simultaneous release of inorganic phosphate from the preformed enolpyruvate nucleophile during C–C bond formation the additions are essentially irreversible and, therefore, these lyases are often referred to as synthases. The mechanistic details of these reactions, however, have yet to be elucidated but it seems obvious that the chances of variation on the part of the nucleophile will be strictly limited. Although the thermodynamic advantage makes these enzymes highly attractive for synthetic applications, none of them is yet commercially available and only few data have been reported concerning the individual specificities towards aldehydic substrates.

phosphoenolpyruvate

4.1 Preparation of Phosphoenolpyruvate

Phosphoenolpyruvate (PEP, **26**), the nucleophile shared by the synthases, is a highly stable compound in aqueous solution under neutral conditions ($t_{1/2} = 10^3$ h at 25 °C) [35]. It is marketed as a highly expensive fine chemical but can be readily prepared in larger than molar quantities from inexpensive pyruvic acid **2** by successive α-bromination to **27**, *Perkow* reaction with trimethyl phosphite, and controlled hydrolysis, after which **26** is precipitated as the stable monopotassium salt in $\sim 50\%$ overall yield [149]. PEP prepared in pure form by this procedure is the preferred source for aldol additions if stringent purity is required because no further components are introduced that may inhibit the catalysts or contaminate the products. Alternatively, **26** can also be produced in situ from commercially available and relatively inexpensive D-3-phosphoglyceric acid **28** by employing the combined action of phosphoglycerate mutase (EC 2.7.5.3) and enolase (EC 4.2.1.11) via the intermediate formation of 2-phosphoglycerate **29** [150]. The overall equilibrium constant of the latter sequence is close to unity but the consumption of **26** in a synthase reaction will drive the overall process.

4.2 N-Acetylneuraminic Acid Synthase

Biosynthesis of sialic acid proceeds in mammals by the phosphorylation of N-acetyl-D-mannosamine **1** to the 6-phosphate, aldol addition with **26** by a specific synthase, and subsequent release of the terminal phosphate ester to give free Neu5Ac **3**. Unlike the mammalian system, some bacteria contain a synthase that acts directly on the unphosphorylated **1** [151]. This latter type of N-acetylneuraminic acid synthase (Neu5Ac synthase or NeuS; EC 4.1.3.19) catalyzes an important step in the biosynthesis of colominic acid, a capsular polysaccharide which is composed of α-2,8-linked sialic acid residues and which is a major virulence factor of serogroup B *Meningococcus* and humanpathogenic *Escherichia coli* of serogroup K1 [80]. The genes encoding the synthase enzymes of *E. coli* K1 [152] and of *Neisseria meningitidis* [153, 154] have been cloned, and the latter enzyme has been isolated and used for the micro-scale synthesis of the parent **3** from **1** and that of its 9-azido analog **30** from N-acetyl-6-azi-do-6-deoxy-D-mannosamine **31** [155].

4.3 2-Keto-3-deoxyoctosonic Acid 8-Phosphate Synthase

The 3-deoxy-D-*manno*-2-octulosonic acid 8-phosphate synthase (KDO synthase or KdoS; EC 4.1.2.16) is an enzyme involved in the biosynthesis of the eight-carbon sugar KDO [63], a constituent of the capsular polysaccharides (K-antigens) and outer membrane lipopolysaccharides (LPS, endotoxin) of Gram-negative bacteria [156], as well as of the cell wall of algae and a variety of plants [157]. In vivo, the enzyme catalyzes the irreversible addition of **26** to D-arabinose 5-phosphate (Ara5P, **32**) to form KDO 8-phosphate **33** [158]. The

KdoS from *E. coli* has been cloned and overexpressed [159–161] to simplify its purification [161–164] and to facilitate mechanistic studies [161, 165].

The purified KdoS enzyme has been utilized both in its soluble [166] or immobilized forms [167] for a one-pot, coupled enzymatic synthesis of **33** in vitro at the multi-gram scale. The protocol employed D-arabinose **10** and two equivalents of **26** as the sole starting materials. The first equivalent of **26** was used for ATP cofactor regeneration in the sugar phosphorylation mediated by the hexokinase/pyruvate kinase couple (EC 2.7.1.1/EC 2.7.1.40), and the second equivalent of **26** in the ensuing C–C bond formation to give **33**. The 2-deoxy and 2-deoxy-2-fluoro analogs of KDO have been prepared by combined enzymatic and chemical procedures [167] in the quest for antibiotics targeted at Gram-negative bacteria [110, 164]. While the substrate tolerance of the microbial KdoS remains yet to be explored, an ambiguity of the enzyme found in pea leaves towards other aldehydes has been reported [168].

4.4 3-Deoxy-D-*arabino*-heptulosonic Acid 7-Phosphate Synthase

The 3-deoxy-D-*arabino*-heptulosonic acid 7-phosphate synthase (DAHP synthase or AroS; EC 4.1.2.15) is an enzyme involved in the shikimate pathway of aromatic amino acid biosynthesis in bacteria and plants where it catalyzes the construction of 3-deoxy-D-*arabino*-heptulosonic acid 7-phosphate **34** from **26** and D-erythrose 4-phosphate (Ery4P, **35**) [169, 170]. The AroS enzyme has been cloned [171–174] and isolated [169, 170, 175–178] from *E. coli* strains. Purification from *Streptomyces rimosus* [179] and gene cloning from *Salmonella typhimurium* [180] and from potato [181, 182] has also been recently achieved. The cytosolic AroS isoenzyme prevalent in higher plants has been described to offer a remarkably broad tolerance for acceptor substrates, including glyoxalate, glycolaldehyde, and most 3–5 carbon sugars or their phosphates [183].

DAHP **34** has been shown to be a valuable precursor to the corresponding DAH phosphonate analog [184], which is a potent inhibitor of dehydroquinate synthase [185]. Because the aldehyde **35** is notorious for its sensitivity and its preparation tedious, an immobilized enzyme system was set up to generate **35** in situ from more readily available and stable precursors (Scheme 5) [184]. In the first step, D-fructose **37** was phosphorylated by hexokinase (EC 2.7.1.1) to the 6-phosphate **38**, from which transketolase (EC 2.2.1.1) removed a C_2-segment by transfer to D-ribose 5-phosphate **39**, thus liberating **35** which then was consumed in the final aldol addition with formation of the required **34**. The twofold function of **26** both as the phosphoryl and aldol donor secured the complete conversion of the starting material. Although this scheme proved highly efficient, the immobilized AroS was found to have a rather dissappointing catalytic lifetime. Larger-scale production of **34** was later found to be simplified when taking recourse to whole cell techniques and using a genetically engineered microorganism that contained a plasmid coding for AroS [186]; thereby substrates and the remaining enzymes were intrinsically provided by regular cell

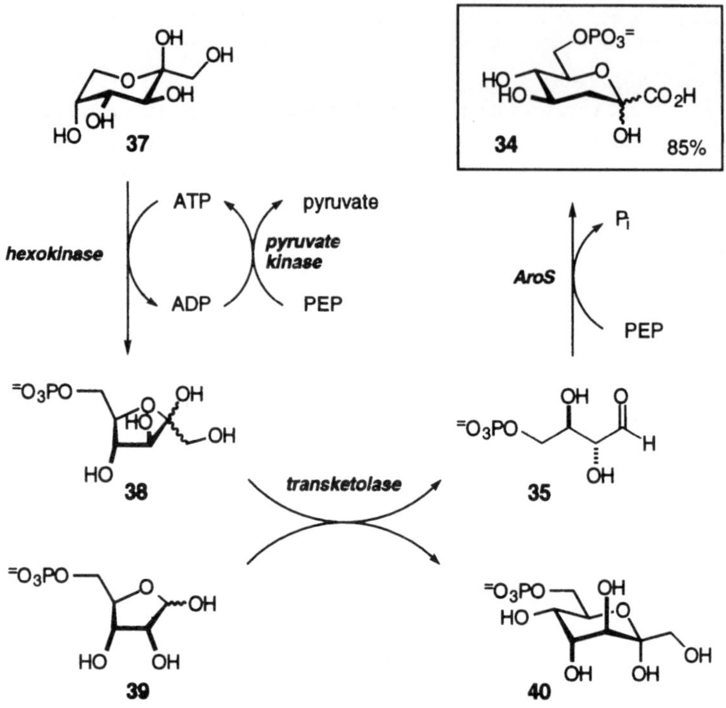

Scheme 5. Combined enzymatic synthesis of 3-deoxy-D-*arabino*-heptulosonic acid 7-phosphate

metabolism. A similar, but even more elaborate approach of constructed microbial metabolism involving the plasmid-based AroS also made feasible a microbial biosynthesis of quinic acid [187] or catechol [188], essentially starting from glucose.

5 Dihydroxyacetone Phosphate Lyases

While the lyases that transfer a pyruvate unit form only a single stereogenic center, the group of dihydroxyacetone-phosphate-(DHAP, **41**)-dependent aldolases create two new asymmetric centers, one each at the termini of the new C–C bond. A particular advantage for synthetic endeavors is the fact that Nature has evolved a full set of four stereochemically-complementary aldolases [189] (Scheme 6) for the retro-aldol cleavage of diastereoisomeric ketose 1-phosphates — D-fructose 1,6-bisphosphate (**42**; FruA), D-tagatose 1,6-bisphosphate (**43**; TagA), L-fuculose 1-phosphate (**44**; FucA), and L-rhamnulose 1-phosphate (**45**) aldolase (RhuA). In the direction of synthesis this formally allows the

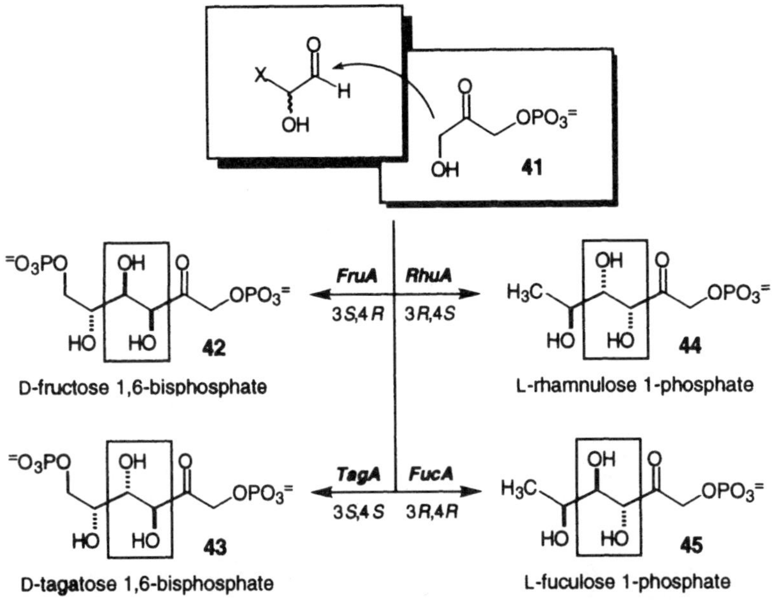

Scheme 6. Stereochemically complementary set of dihydroxyacetone-phosphate-dependent aldolases

deliberate preparation of any one of the possible four diastereomeric aldol adducts in a building block fashion [189, 190] by simply choosing the appropriate enzyme and starting materials for a full control over constitution and absolute configuration of the desired product (cf. Sect. 7).

The Schiff-base-forming types (class I) are known only for the two former aldolases (FruA, TagA), which are found usually in mammalian or (as an exception) in specific microbial organisms, whereas the Zn^{2+}-dependent type (class II) comprises all four DHAP aldolases which are commonly found in bacteria [43]. Typically, type I FruA enzymes are tetrameric proteins composed of subunits of ~40 kDa [191, 192], while the type II FruA are dimers of ~39 kDa subunits [193]. RhuA and FucA enzymes are homotetrameric with a subunit molecular weight of ~25 kDa and ~30 kDa respectively [194, 195].

Representatives of all kinds have been explored for synthetic applications while mechanistic investigations were mainly focussed on the distinct FruA enzymes isolated from rabbit muscle [196] and yeast [197, 198]. For mechanistic reasons, all DHAP aldolases appear to be highly specific for the donor component DHAP [199], and only a few isosteric replacements of the ester oxygen for sulfur (46), nitrogen (47), or methylene carbon (48) were found to be tolerable in preparative experiments (Fig. 7) [200, 201]. Earlier assay results [202] that had indicated activity also for a racemic methyl-branched DHAP analog 53 are now considered to be artefactual [203]. Dihydroxyacetone sulfate 50 has been shown to be covalently bound via Schiff base formation, but apparently no α-deprotonation occurred as neither H/D-exchange nor C–C

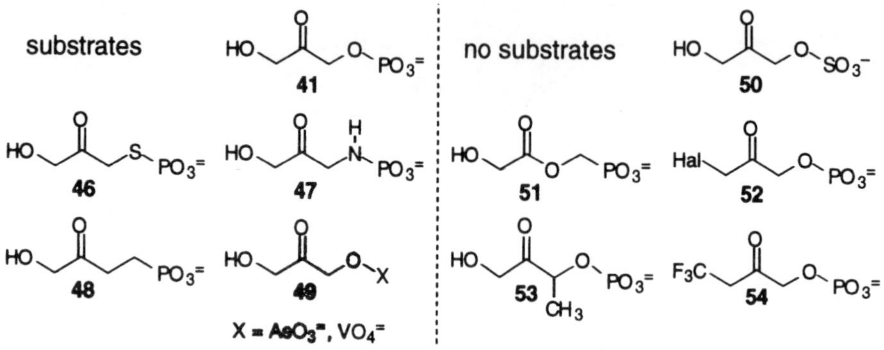

Fig. 7. Substrate quality of dihydroxyacetone phosphate analogs

bond formation was induced [199]. No replacement of the hydroxy function for other electron-withdrawing substituents is tolerated [202, 204]. Instead, haloacetol phosphates **52** [204] and the related trifluoromethyl analog **54** [205] cause rapid "suicide" enzyme inactivation presumably due to alkylation of active site nucleophiles.

Except for the type II FruA enzymes (see below), all DHAP aldolases accept a vast range of differently-substituted or unsubstituted aliphatic aldehydes at synthetically useful rates as the acceptor component [202]. Limitations are met only with nonpolar tertiary (pivaldehyde), α, β-unsaturated, and aromatic aldehydes which apparently are not converted by any one of the DHAP aldolases. An exception to the last criterion are heteroaromatic carbaldehydes containing a nucleophilic ring nitrogen (e.g. pyridine derivatives **66**) which are clearly bound in an N-protonated state, according to a recent determination of pH influence on reaction rates [206]. Generic aliphatic α,ω-dialdehydes have assayed positive but no distinct product could be isolated in preparative scale experiments so far [207]; rather, dialdehydes lead to rapid deactivation of the catalysts due to cross linking (for an exception, see Sect. 7.1). The highest conversion rates, diastereoselectivities, and yields are generally achieved with 2- or 3-hydroxyaldehydes. Higher yields in the latter cases usually derive from the fact that the products readily cyclize in aqueous solution to form thermodynamically stable furanoid or pyranoid rings (Scheme 7). As a compromise between the instability of DHAP, but higher enzyme activity at pH values above 7, reactions are preferably run at slightly acidic pH. From the ketose 1-phosphates produced in this way, the corresponding non-phosphorylated compounds may be readily liberated by chemical or, preferably, by mild enzymatic hydrolysis. Depending on product stability either an inexpensive alkaline phosphatase (EC 3.1.3.1) at pH 7.5–9.0 is recommended [208], while base labile compounds require a more expensive acid phosphatase (EC 3.1.3.2) and working at pH 4.5–6.0 [202].

As a rule, the Zn^{2+}-dependent microbial aldolases of type II are much more stable in solution than their mammalian counterparts (type I) and have half-lives

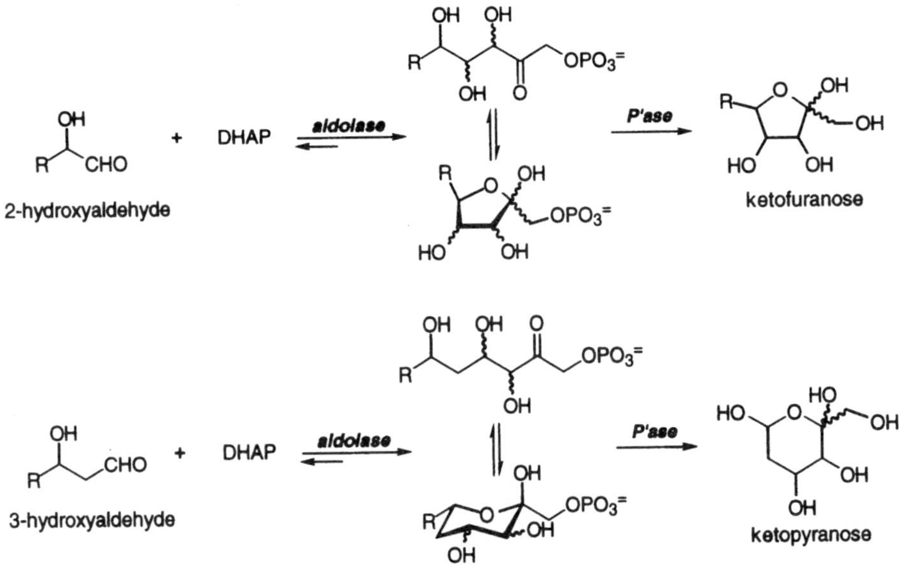

Scheme 7. Formation of stable ketofuranoses and ketopyranoses from 2- or 3-hydroxyaldehydes

of several weeks or even months [195, 209] as compared to only a few days [202]. A notable exception is the ~ 33 kDa, monomeric class I FruA from *Staphylococcus aureus*, which shows unusual thermal and pH stability [210]. For experiments with nonpolar substrates with low solubility in water, the metalloproteins tolerate relatively high proportions (up to 50%) of organic cosolvents (such as lower aliphatic alcohols, acetonitrile, DMF or, preferably, DMSO) and have enzyme half-lives that are still in the range of several weeks (Fig. 8) [195]. Stability of type II aldolases is closely linked to the presence of low concentrations (~ 0.5 mM) of metal ions because of their structural role in confining the active site which is located at the subunit interface of the oligomeric active catalysts [194, 211]. Attempts at catalyst immobilization have so far proved either inefficient or ineffective. Although rabbit muscle FruA could be immobilized on microcarrier beads [56], the minor increase in stability could not outweigh the loss of activity incurred during covalent attachment; severe inactivation has so far also accompanied attempts to immobilize the type II aldolases such as FruA, RhuA, or FucA [212]. Presumably this is due to a distortion of mobile enzyme domains such as the flexible C-terminus, which for some aldolases is known to participate in catalytic events [213–216]. Very good operational stabilization is expected from cross linked enzyme crystals (CLEC) which have been prepared for rabbit muscle FruA [57] and the *E. coli* RhuA enzymes [58], although data for long term stability, possible diffusion limitations, or restrictions on substrate structure and size have not yet been determined.

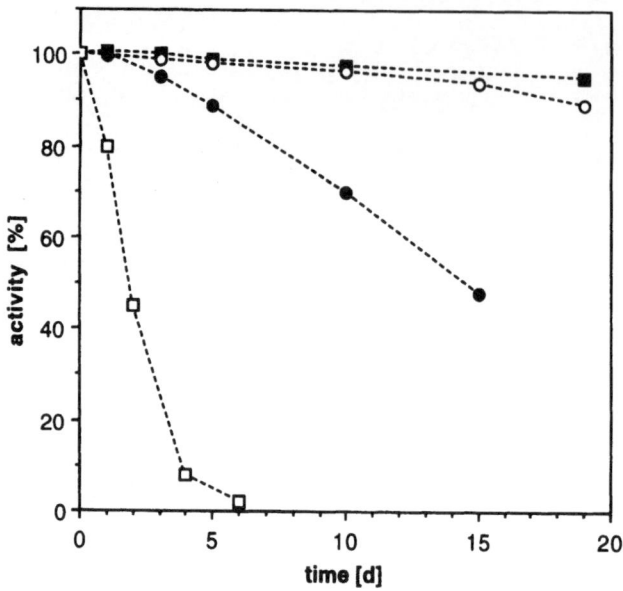

Fig. 8. Stability of the rhamnulose 1-phosphate aldolase from *Escherichia coli* (RhuA) vs. that of the fructose 1,6-bisphosphate aldolase from rabbit muscle (FruA) in phosphate buffer (pH 7.2; 25°C; ca. 1 U ml^{-1}) a) ■ RhuA; b) ○ RhuA, 30% EtOH; c) ● RhuA, 50% DMSO; d) □ FruA

For any one of the DHAP aldolases, the absolute configuration at the newly created stereocenter at C-3 is invariably conserved upon reaction with any electrophile, apparently for mechanistic reasons [199]; no exceptions are known so far. For the stereogenic center at C-4, the relative positioning of the aldehyde carbonyl in the transition state, and thus the relative configuration in the product, usually follows closely that of the natural substrates. Depending on the nature of the enzyme used and on the pattern of substitution present in the aldehydic component, a distinct fraction of the 4-epimeric diastereomers may also be observed which is presumably the result of incorrect binding of the respective aldehyde (cf. Sect. 3.1).

5.1 Preparation of Dihydroxyacetone Phosphate

Owing to the narrow specificity of the DHAP aldolases for the donor substrate DHAP (**41**), direct access to this essential compound is vital to the development of synthetic applications. Commercial offers of the compound, however, are prohibitively expensive for preparative-scale applications. A further problem is that **41** is relatively unstable in solution and, particularly at elevated pH values, readily decomposes according to an Elcb pathway via an enediol intermediate into methylglyoxal **55** and inorganic phosphate [217–219], both of which are inhibitors of the aldolases (e.g. < 10% RhuA activity at 50 mM P_i) [220].

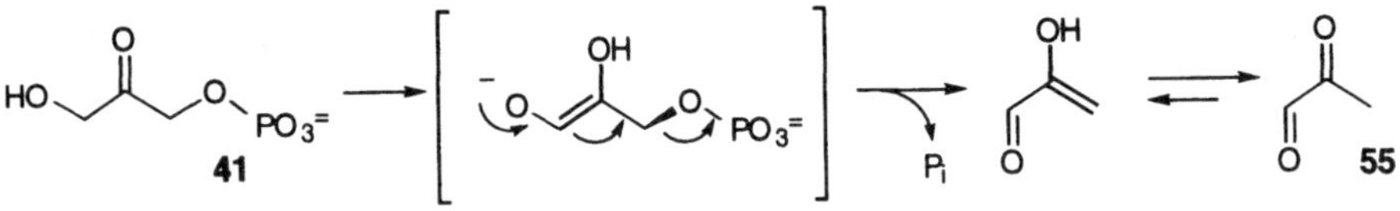

Hence, several protocols have been developed for its synthesis by phosphorylation of dihydroxyacetone **56**, which can be achieved either by a multistep chemical route or directly by an enzymatic method (Scheme 8). Chemical phosphorylations start from dimeric diethyl acetal **57** [221] in which the unprotected hydroxyl groups are esterified either by using phosphoroxy chloride followed by hydrolytic work-up [222], by phosphitylation with dibenzyl N,N-diethyl phosphoramidite and H_2O_2 oxidation to the phosphate stage [223], or by treatment with diphenyl chlorophosphate [224, 225]. In the two latter cases the sequence is completed by hydrogenolytic deprotection of the resulting **58** to **41**. After acidic hydrolysis and a number of incremental improvements, free DHAP may now be obtained in up to 60% overall yield [225]. A certain disadvantage remains in the competing phosphate ester hydrolysis during the acetal deprotection because of the formation of considerable quantities of inorganic phosphate (⩾ 20%), which is inhibitory to the aldolases, and of dihydroxyacetone, which may interfere with product purification. To some extent, this difficulty may be ameliorated by employment of the corresponding dimethyl acetals **57/58** (R=Me) whereupon only ⩽ 5% of ester hydrolysis occurs during liberation of **41** [195].

Alternatively, enzymatic phosphoryl transfer to **56** is achieved by using ATP-dependent kinases. Glycerol kinase (EC 2.7.1.30) accepts **56** as a substrate analog [226], plausibly with the carbonyl group hydrated. The reaction, including a cofactor recycling scheme using PEP (or acetyl phosphate) as the ultimate phosphoryl donor, has been scaled up efficiently to the mole scale without difficulty [227]. The disadvantages are the need for a separate preparation of the phosphate donor, and complications arising during the purification of aldol adducts from the side products released stoichiometrically during ATP regeneration (pyruvate, acetate). A specific dihydroxyacetone kinase (EC 2.7.1.29)

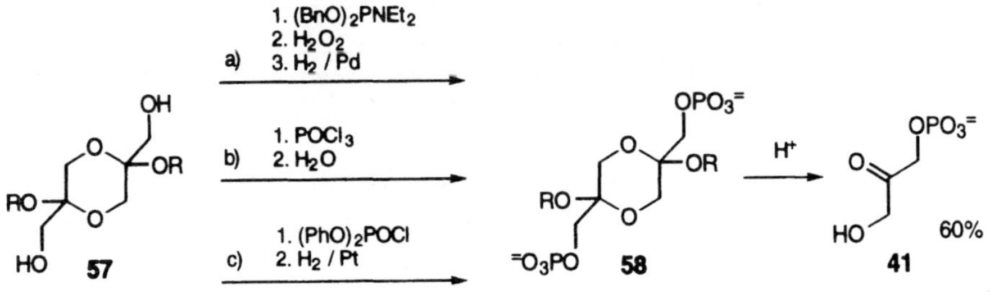

Scheme 8. Selective chemical phosphorylation of dimeric dihydroxyacetone

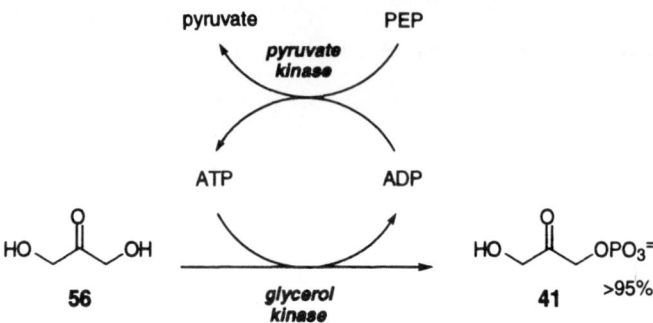

Scheme 9. Enzymatic phosphorylation of dihydroxyacetone with in-situ cofactor regeneration

from methylotrophic yeast has also been applied for the phosphorylation of **56** by ATP in a fed-batch reaction to produce DHAP in up to 175 mM concentration [165].

Because of its sensitive nature, however, **41** is not only best generated enzymatically but is also consumed in situ by enzymic aldol additions that avoid the build-up of high stationary concentrations. The most elegant, efficient, and convenient method is the in situ formation of two equivalents of **41** from commercially available fructose 1,6-bisphosphate (FBP, **42**) by a combination of FruA and triose phosphate isomerase (EC 5.3.1.1; see Scheme 10, inset) [202]. The technique dates back to the discovery of the FruA enzyme which at that time — 60 years ago — had already been utilized to form ketose 1-phosphates from **42** and simple aldehydes [228]. The equilibria of both the component aldol and the isomerization reactions are each in favor of the respective kinetically more stable compound [229], i.e. **42** and **41**. The disadvantage of the relatively high costs of pure **42** as the starting material has been circumvented by the design of a highly integrated, multienzymatic scheme (Scheme 10) that allows the efficacious in situ preparation of two DHAP equivalents from inexpensive glucose **59** or fructose **37**, and that of four equivalents from sucrose **60**, respectively, via the in situ formation of **42** by using a set of up to seven commercially available and inexpensive enzymes [230]. The sequence mimics parallel branches of glycolysis pathways in vitro and because of its unusual complexity can be designated as an "artificial metabolism". Problems arising from mutual inhibition between participating enzymes and "metabolites" formed en route had been identified as being due to the commercial mammalian phosphofructo-kinase (EC 2.7.1.11) [231], but could be solved by replacing the bottleneck catalyst by a less-regulated isoenzyme [232] which was isolated from a genetically-engineered *E. coli* strain [233]. This "artificial metabolism" has been applied to the preparation of a number of FruA adducts (*vide infra*) and has proved to work very successfully even with very poor substrate analogs [230]. Yields of aldol adducts were generally close to those obtained by starting from pure commercial **42**.

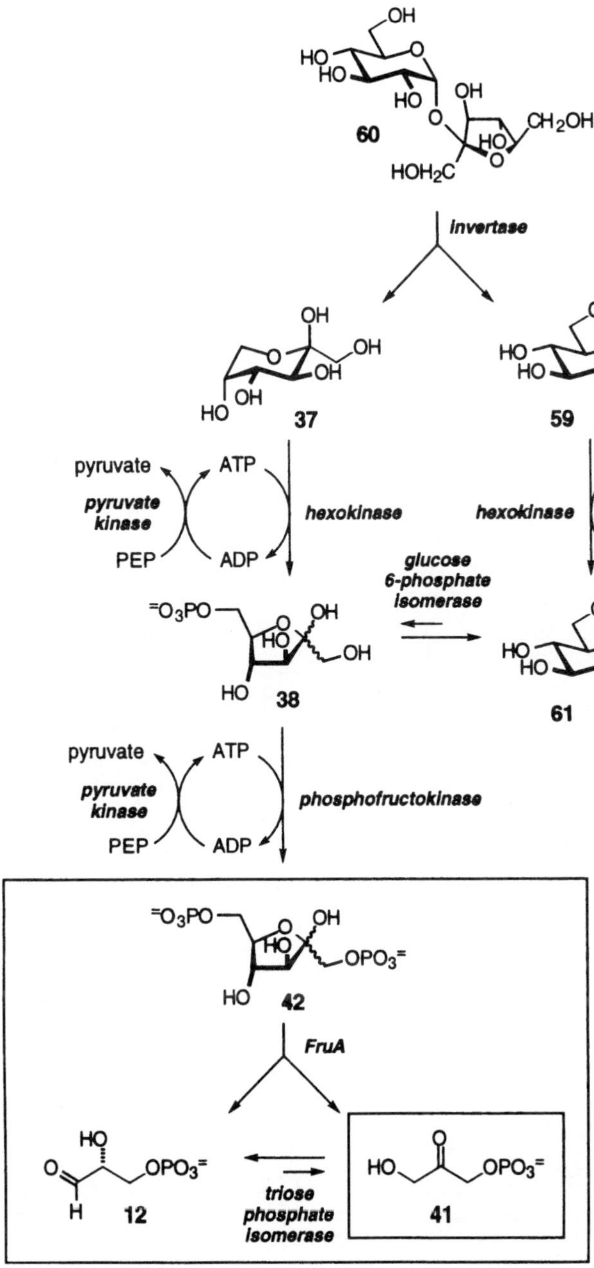

Scheme 10. 'Artificial metabolism' for the in situ preparation of DHAP from regenerable carbohydrate resources

The method based on **42**, however, may pose problems with aldehydes carrying anionic charges (e.g. carboxylates or phosphates) because of thermodynamically less favorable equilibria and difficulties arising during product purification due to the similar ionization states of starting material, possible contaminants, and products [202]. Apparently, because of the presence of a FruA enzyme, it is also restricted to applications involving an identical FruA stereopreference in the synthesis direction. This restriction was recently lifted when it was discovered [234] that the type II FruA of *E. coli* (much like its yeast equivalent) — contrary to perpetuated assertions [209, 235–237] of broad synthetic applicability — has a strict substrate specificity for D-glyceraldehyde 3-phosphate **12** and thus may be used for cleavage of **42** without interfering with synthetic reactions that are catalyzed by other aldolases of distinct stereoselectivity (Scheme 11). Such metabolically engineered artificial metabolisms, encompassing up to 8 different enzymes working in concert, have been shown to produce diastereomerically pure *erythro* products **62** when supplemented with the *erythro* selective FucA from *E. coli* [234].

Alternatively, DHAP **41** can be formed from glycerol **64** as another inexpensive, regenerative source via L-glycerol phosphate (G3P, **65**) [208]. The successive phosphorylation and oxidation was effected by employing a combination of glycerol kinase (EC 2.7.1.30) and glycerol phosphate dehydrogenase (EC 1.1.1.8). For the indispensable double cofactor regeneration (Scheme 12) an integrated closed-loop system was developed which employs phosphoenolpyruvate **26** as the sacrificial reagent for sequential ATP and NAD$^+$ recycling steps by pyruvate kinase (EC 2.7.1.40) and L-lactate dehydrogenase (EC 1.1.1.27). Because of the very similar redox potentials of **2** ($E'_0 = -0.185$ V) and **41** ($E'_0 = -0.192$ V) [35], this system was synthetically useful only when it was coupled to exergonic aldol additions, such as in the RhuA-catalyzed synthesis of several stable, cyclic sugar derivatives [208].

The latest, and most advanced, technique uses an inexpensive flavine-dependent glycerol phosphate oxidase (GPO, EC 1.1.3.21), found in several microorganisms (Scheme 13), for air oxidation of L-glycerol 3-phosphate **65** to generate DHAP practically quantitatively and in high chemical purity [200]. A separate cofactor regeneration step has become obsolete because the reduced

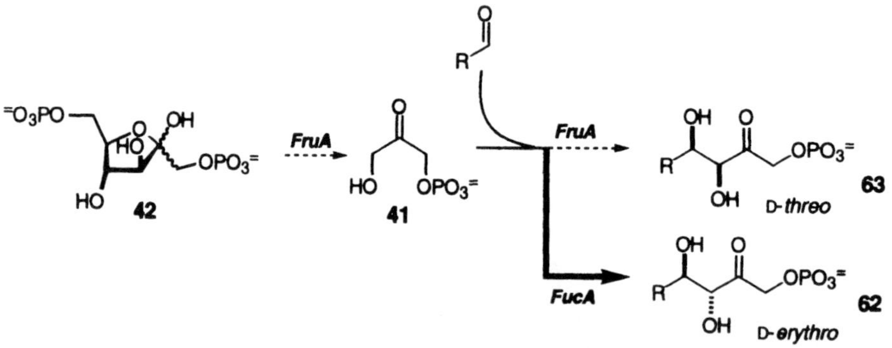

Scheme 11. Metabolic engineering of an 'artificial metabolism' for stereoselective C–C bond formation by selecting appropriate catalysts

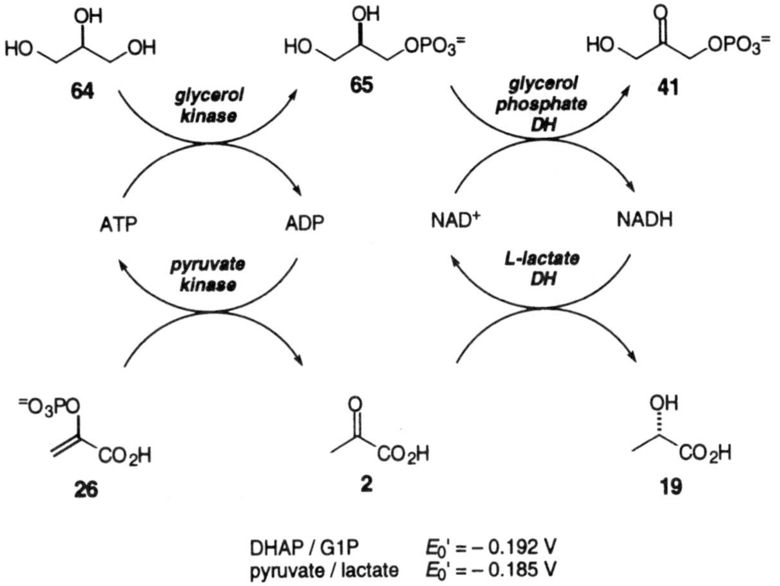
64
glycerol kinase
65
glycerol phosphate DH
41
ATP
ADP
NAD+
NADH
pyruvate kinase
L-lactate DH
26
2
19
DHAP / G1P $E_0' = -0.192$ V
pyruvate / lactate $E_0' = -0.185$ V

Scheme 12. 'Artificial metabolism' for closed-loop ATP/NAD$^+$-cofactor regeneration, applied to the multienzymatic in-situ preparation of DHAP from glycerol

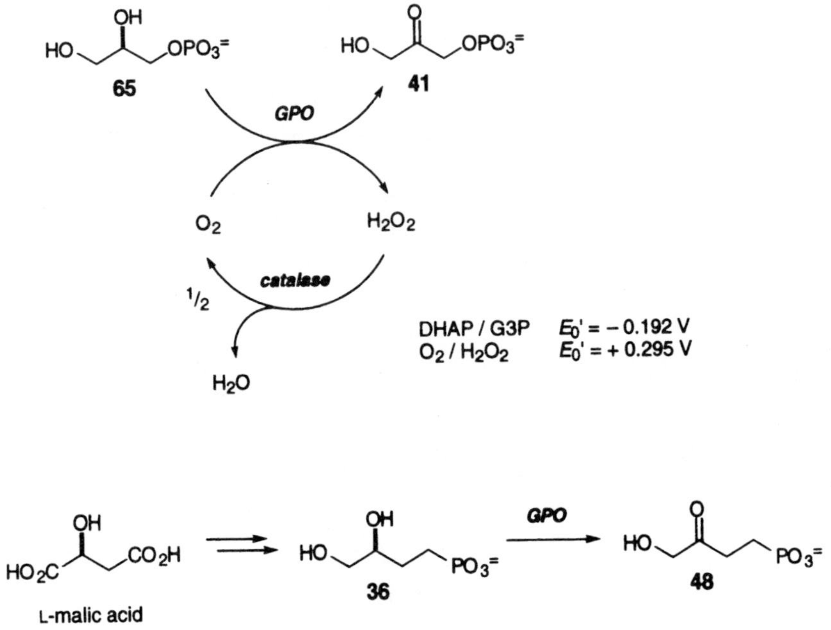
65
41
GPO
O_2
H_2O_2
catalase
1/2
H_2O
DHAP / G3P $E_0' = -0.192$ V
O_2 / H_2O_2 $E_0' = +0.295$ V
L-malic acid
36
GPO
48

Scheme 13. Enzymatic oxidation of glycerol phosphate for the in situ preparation of DHAP, and formation of an isosteric phosphonate analog

cofactor $FAD(H_2)$ remains tightly bound to the oxidase and is rapidly re-oxidized by elemental oxygen. The hydrogen peroxide thereby liberated is decomposed by added catalase (EC 1.11.1.6); as the flow chart suggests, H_2O_2 can be used to sustain oxygenation. Although the GPO is specific for the L-enantiomer [238], cheap crude glycerol phosphate racemate can be used because regio- and stereoisomers are no inhibitors [200]. Since all the DHAP aldolases tested proved insensitive to oxygenated solutions, oxidative genera-tion of DHAP and its consumption in aldol reactions could be smoothly coupled. When compared to the established procedures, the GPO method usually furnished adducts of higher purity and in at least equal or higher yields. An interesting observation is that by using the cyclohexylammonium salt of **65**, specific aldol adducts can be induced to crystallize directly from the crude product solution in high yield and purity; this significantly simplifies the work up. Furthermore, the GPO procedure proved adaptable to the synthesis of DHAP analogs, modified at the phosphate group (i.e. **46**, **47**, and **48**), which are substrates of the aldolases. From the enantiomerically pure dihydroxybutyl-phosphonate **36**, an isosteric mimic of **65** which is readily obtained from natural L-malic acid [239], non-hydrolyzable sugar phosphonates can thus be syn-thesized in a one pot operation.

Remarkably, a mixture of **56** and inorganic arsenate can replace **41** due to the spontaneous, transient formation of a monoarsenate ester **49a**, which is recognized by the aldolase as a DHAP mimic (Fig. 7) [240]. Since the arsenate ester formation is the rate-determining step, these aldol reactions are character-ized by an apparent irreversibility that is synthetically useful. However, this approach suffers from generally low reaction rates due to the low stationary concentration of **49a** in solution, from the high toxicity of inorganic arsenate, especially at the high levels ($\geqslant 0.5$ M) required for efficient conversions to occur, and from problems in product isolation. Inorganic vanadate also spontaneously forms [241, 242] the corresponding vanadate ester **49b**, but has to be modulated in its oxidation potential by appropriate buffer bases (e.g. imidazole [243]) to prevent oxidation of dihydroxyacetone [234]. Since the rate-determining es-terification is faster by several orders of magnitude with vanadate as compared

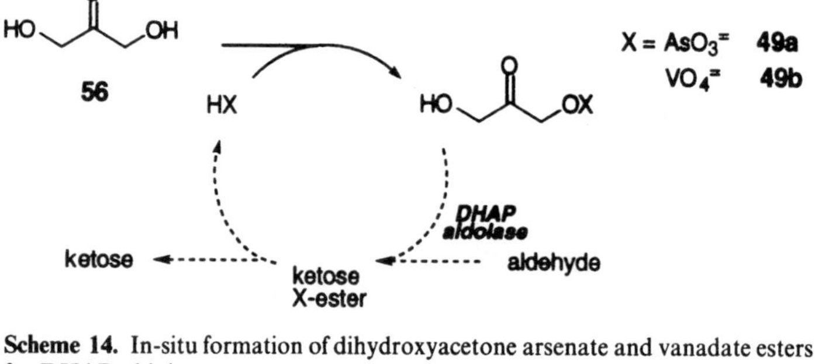

Scheme 14. In-situ formation of dihydroxyacetone arsenate and vanadate esters as donor substrates for DHAP aldolases

to arsenate [242], very low vanadate concentrations ($\leqslant 0.1$ mM) suffice to sustain rapid catalytic turnover. For reasons yet to be determined, so far only the RhuA of *E. coli* was capable of accepting the vanadate mimic **49b** [234].

5.2 D-Fructose 1,6-Bisphosphate Aldolase

In vivo, the D-fructose 1,6-bisphosphate aldolase (FruA; EC 4.1.2.13) catalyzes the pivotal reaction of the glycolysis pathway: the equilibrium addition of **41** to D-glyceraldehyde 3-phosphate (GA3P, **12**) to give D-fructose 1,6-bisphosphate (**42**) [43]. The equilibrium constant of 10^4 M^{-1} strongly favors synthesis [229].

Since its discovery, the class I FruA isolated from rabbit muscle [196] has been the most extensively investigated aldolase for mechanistic, structural, and preparative purposes. A variety of class I or class II subgroup FruA enzymes of microbial [197, 198, 210, 244–254], algal [255–257], plant [258–262], insect [263, 264], parasite [265, 266], amphibia [267], fish [268], and mammalian tissue [191, 269–274] origin have also been isolated and characterized, and a number of their genes have been cloned [266, 275–278] for protein overexpression [209, 279, 280]. Because of their ubiquitous occurrence and central role in metabolism, sequence homologies (including conservation of active site residues) as determined for a multitude of FruA enzymes are of special concern in phylogenetic studies of species evolution [255]. High structural similarities became also evident from X-ray structure determinations of the class I FruA proteins from rabbit (2.7 Å) [281] and human muscle (2.0 Å) [282, 283], and from *Drosophila melanogaster* (2.5 Å) [284]. The molecular architecture of the enzyme subunit corresponds to an eight stranded α/β-barrel structure which is typical for glycolytic enzymes. The active-site cleft is located in the center of the β-barrel with the reactive Lys229 projecting towards the C-terminal opening of the pocket (Fig. 10). Several other basic residues seem to cluster at opposing ends of the pocket (Lys107, Lys146, and Arg148, *versus* Lys41, Arg42, and Arg303), possibly to bind the individual phosphate moieties of the natural substrate. A comparison of the X-ray structures of the rabbit (Fig. 9A) and *Drosophila melanogaster* enzymes (Fig. 9B) possibly visualizes the function of the

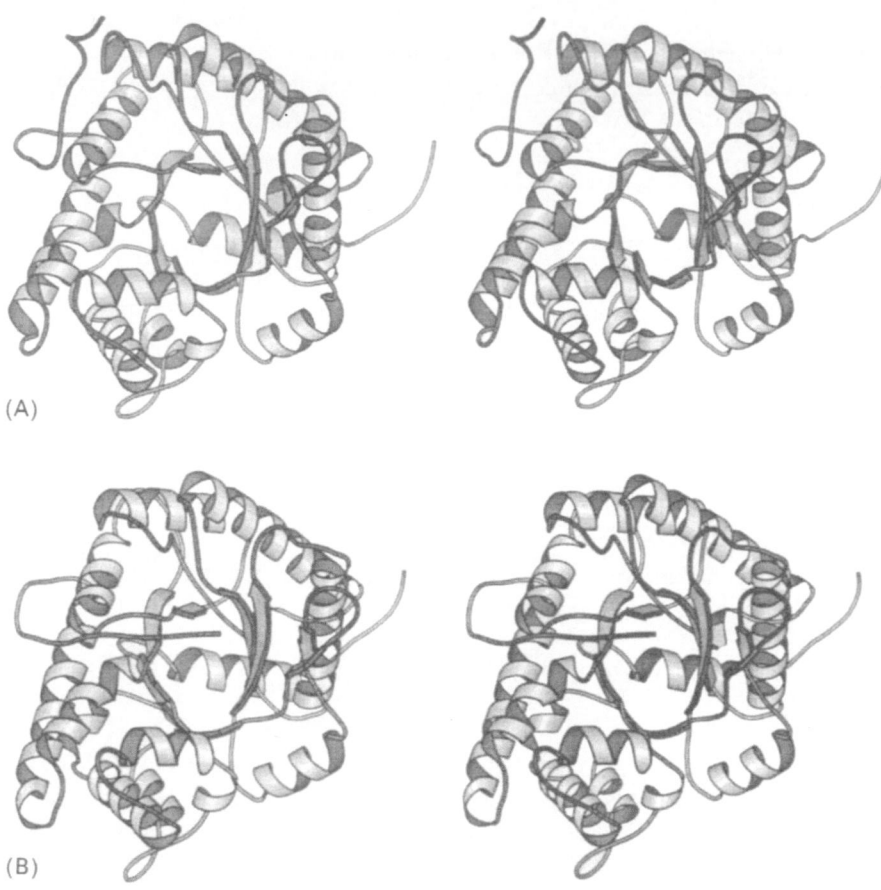

Fig. 9. Stereo ribbon plots [76] of FruA subunits viewed down the β-barrel axis from their carboxy-terminal end to illustrate the flexibility of the C-terminal chain as proposed for the catalytic function. A) Crystal structure of the FruA from rabbit muscle; B) crystal structure of the FruA from *Drosophila melanogaster*

carboxy terminus, which determines the substrate specificity of the various FruA enzymes and which is known to be involved in the catalytic cycle [213–216]. In the rabbit enzyme structure [281], this arm is sharply turned at a flexible hinge region around residue 350 and projects out into solvent, allowing substrate to freely enter the open active site. In the *Drosophila* enzyme [284] this polypeptide segment covers the active-site cleft and, in effect, mediates substrate binding and solvent access.

Of the several aldolases that are now commercially available, the rabbit muscle enzyme with a useful specific activity of ~ 20 U mg^{-1} still remains the most cost-efficient catalyst for preparative work, although the reported higher stability of the *Staphylococcus carnosus* enzyme is a challenge [252, 285]. Literally hundreds of aldehydes have so far been tested successfully by enzymatic

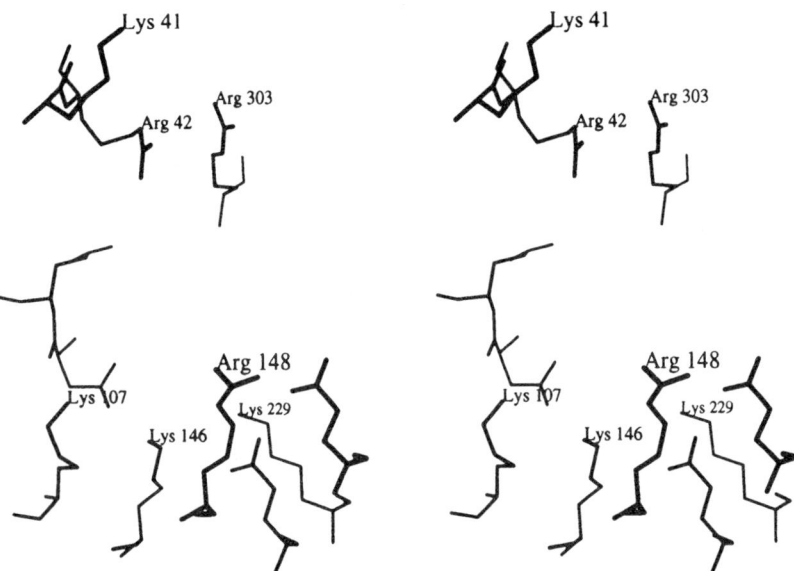

Fig. 10. Stereo view of the active site of the FruA from *Drosophila melanogaster* showing the environment of the Schiff-base-forming Lys229 residue. Clusters of basic residues (L41, R42, R303 and L107, L146, R148) are proposed to be involved in the binding of the two phosphate ester groups

assay and preparative experiments as a replacement for **12** in rabbit muscle FruA-catalyzed aldol additions [21], and most of the corresponding aldol products have been isolated and characterized. A compilation of selected typical substrates and their reaction products is provided in Table 4, and further examples are indicated in the ensuing Schemes as well as in Sect. 7. On the other hand, the class II FruA enzymes of microbial origin — contrary to literature reports [209, 235] — seem to have evolved to a very high substrate specificity even for the aldehydic substrate as demonstrated by the lack of the *E. coli* and yeast enzymes for acceptance of nonphosphorylated substrate analogs (< 1% activity) [286].

In each successful case, the absolute configuration of the resulting vicinal diols has been shown to follow precisely that of the natural fructose stereochemistry, i.e. to be always D-*threo* or (3S, 4R) [21]. Particularly suitable in this respect are aldehydes which carry a hydroxyl function for reasons of higher affinity to and correct recognition by the biocatalyst, and because of the formation of thermodynamically stable cyclic products. For structural resemblance to the natural substrate, phosphorylated aldehydes are better substrates than unphosphorylated ones [202], and even sugar phosphates react (e.g. **32, 35, 61**) [297], probably because of a high binding affinity. More recently, however, it has been discovered [298] that some particular aldehydes containing a (basic) nitrogen functionality, e.g. pyridine carbaldehydes **66** or diethylaminoacetaldehyde, may give rise to up to one third of unexpected diastereomers which are epimeric at that terminus of the newly-formed bond which

Table 4. Substrate tolerance of fructose 1,6-bisphosphate aldolase

R	Rel. rate [%]	Yield [%]	Ref.
D-CHOH-CH$_2$OPO$_3^=$	100	95	[200, 202, 287]
H	105	–	[202]
CH$_3$	120	–	[202, 228]
CH$_2$Cl	340	50	[202, 288]
CH$_2$-CH$_3$	105	73	[202]
CH$_2$-CH$_2$SC$_6$H$_5$	–	33	[289]
CH$_2$-CH$_2$-COOH	–	81	[200]
CH$_2$OCH$_2$C$_6$H$_5$	25	75	[202]
CH(OC$_2$H$_5$)$_2$	–	60	[290]
D-CH(OCH$_3$)-CH$_2$OH	22	56	[202]
CH$_2$OH	33	84	[200, 202]
D-CHOH-CH$_3$	10	87	[202, 291, 292]
L-CHOH-CH$_3$	10	80	[202, 291, 293]
DL-CHOH-C$_2$H$_5$	10	82	[202, 291]
CH$_2$-CH$_2$OH	–	83	[230, 235, 292]
CH$_2$-C(CH$_3$)$_2$OH	–	50	[294]
DL-CHOH-CH$_2$F	–	95	[292]
DL-CHOH-CH$_2$Cl	–	90	[235, 295]
DL-CHOH-CH$_2$-CH=CH$_2$	–	85	[235, 296]

is derived from the aldehyde (i.e. 3S, 4S configuration). Diastereoselectivity with acetaldehyde, a case where the aldolase has to differentiate methyl vs. hydrogen substituents of the electrophile, at de $\sim$98 is highly in favor of si-face attack, with traces ($\sim$1%) of the *erythro* diastereomer only being detectable by in-situ high-frequency ^{1}H NMR spectroscopy [220].

The rabbit FruA discriminates the enantiomers of its natural substrate with a 20:1 preference for D-GA3P (**12**) over its L-antipode [202]. Assistance from anionic binding was revealed by a study on a homologous series of carboxylated 2-hydroxyaldehydes which showed optimum enantioselectivity when the distance of the charged group equaled that of **12** (Scheme 15, Fig. 11) [299]. The resolution of racemic substrates is not, however, generally useful since the kinetic enantioselectivity for nonionic aldehydes is rather low [202]. 3-Azido substituents (**69**) can lead to an up to 9-fold preference of enantiomers in kinetically controlled experiments [300] while hydroxyl (**70**; preference for the

n	73 : 74
1	3 : 1
2	14 : 1
3	10 : 1
5	7 : 1

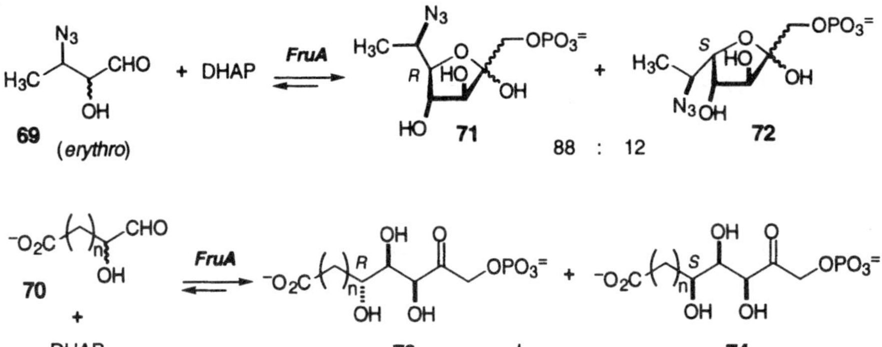

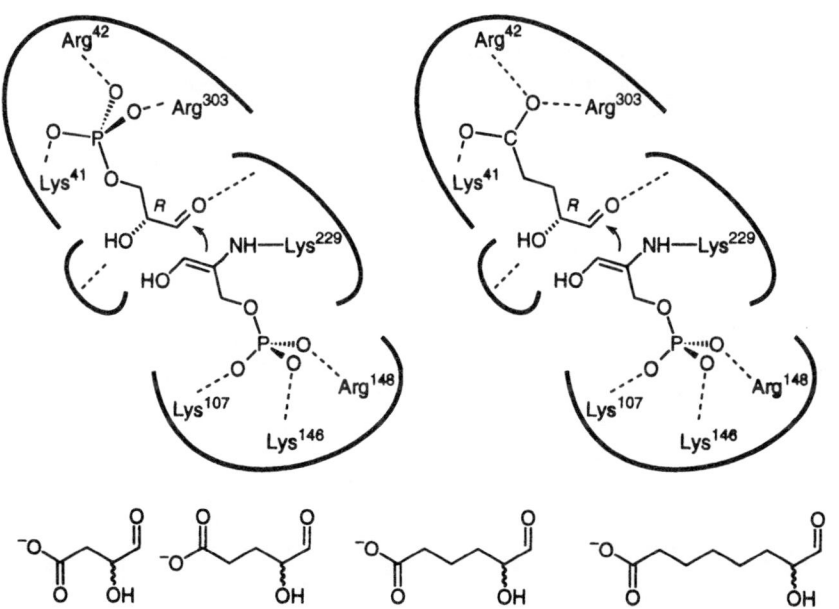

Scheme 15. Examples of the kinetic enantioselectivity of the FruA from rabbit muscle

Fig. 11. Schematic representations for substrate binding by the FruA from rabbit muscle to rationalize the origin of enantioselectivity observed with α-hydroxyaldehydes carrying a distant carboxylate function

D-antipodes) and derived functions or larger distance of chiral centers rarely exceed statistical diastereomer formation.

Thermodynamic control of diastereomer formation under fully equilibrating conditions can be utilized with racemic 2-[235, 295] or 3-hydroxylated

[202, 235, 294,296] aldehydes to achieve a high level of enantiomer differentiation because the products adopt a cyclic ring form (Scheme 16). Isomers having the discriminating substituent in a *trans* or equatorial position will predominate with up to 3.5-fold [235, 295] or 33-fold [202, 235, 294, 296] selectivities for furanose- or pyranose-type products, respectively, because of reduced vicinal or 1,3-diaxial repulsions. Thus, in FruA-catalyzed reactions (2R)- and (3S)- configurated hydroxyaldehydes **75** are the preferred substrates. Similarly, 2-alkylated aldehydes **76** can be resolved because of the high preference of the alkyl group for an equatorial position (**77**) [294]. An all-equatorial arrangement of substituents in the more stable product often also enhances the proclivity of the sugar phosphate salts to crystallization [200, 296]. The remaining mixture of diastereomers **78/79** can then be re-subjected to further equilibration in order to maximize the yield of the preferred isomer **78**. This technique has recently found an application in a novel approach to the de novo synthesis of 4,6-dideoxy sugars such as 4-deoxy-L-fucose **81** and its trifluoromethylated analog **82** (Scheme 17) [296]. The thermodynamic origin of stereoselectivity has been proven by isolating the unfavored (6R)-diastereomer **79**, synthesized separately from enantiomerically pure (R)-3-hydroxybutyraldehyde (R)-**84** under FruA catalysis, and subjecting it to a treatment with excess of racemic aldehyde **84** and FruA under equilibrating conditions [202]. Indeed, this led to an equilibrium mixture identical with that generated directly from racemic aldehyde in favor of the opposite (6S)-diastereomer **78** (3:97 ratio).

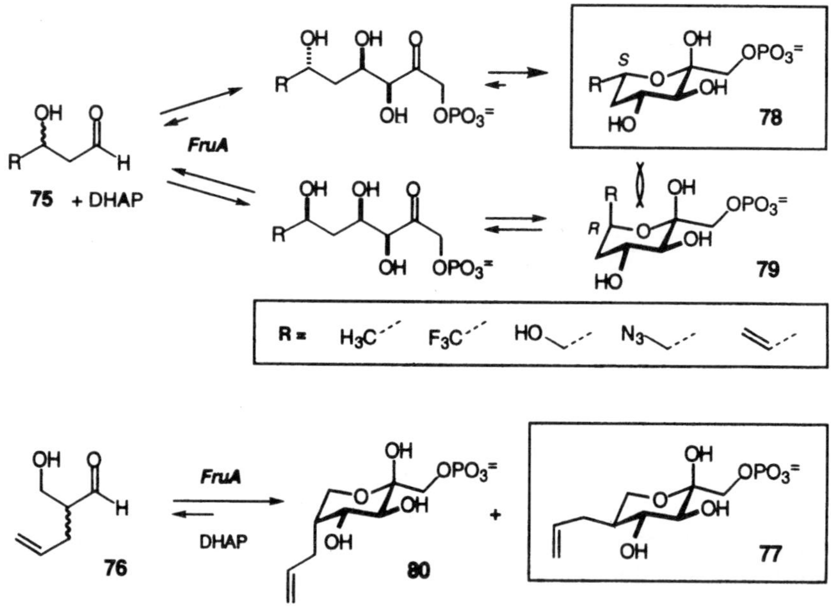

Scheme 16. Thermodynamically controlled FruA-catalyzed aldol additions to 3-hydroxyaldehydes

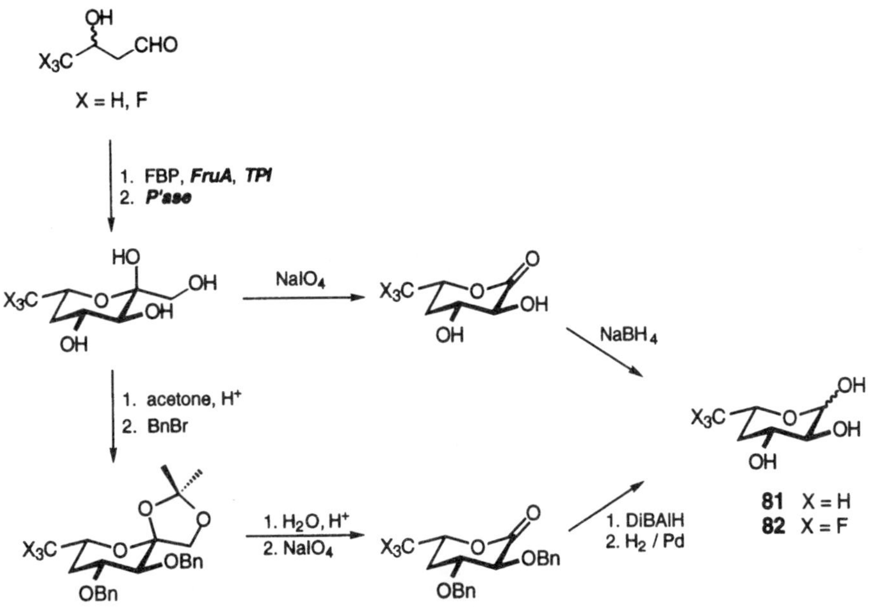

Scheme 17. Synthesis of L-fucose derivatives based on thermodynamic preference of aldolizations with 3-hydroxyaldehydes

Long equilibration of reactions with 3-hydroxybutyraldehyde **84** under FruA catalysis gave rise to a third diastereomeric compound which ultimately accumulated besides the favored **78** (Scheme 18). The former compound could also be obtained selectively by separate equilibration of the unfavored (6R)-diastereomer **79** under FruA catalysis and was identified as the 4-epimeric diastereoisomer **83** having a vicinal *cis* relationship of hydroxyl substituents at the newly created C–C bond (i.e. L-*erythro* or TagA stereochemistry) [301]. Independent proof of stereochemistry was obtained by simply subjecting the pure (6S)-diastereomer **78** to a treatment with a FruA-FucA aldolase mixture [296]. Under these conditions, an equilibrium mixture of **78/85** ≈ 3:2 was generated, reflecting the stabilities expected from increment calculations [302]. The spectra of the FucA product **85** fully matched those of **83**, thus proving the identical relative 3,4-*erythro* connectivity but enantiomeric nature. The generation of an *erythro* product even from simple substrates attests that FruA can no longer be regarded as "stereospecific" although a very high stereoselectivity (normally higher than with all or most other aldolases investigated so far) will routinely be observed in the majority of applications. The very slow formation of a *cis* diol, which for complete conversion requires long reaction times and a relatively large amount of catalyst, indicates that the reaction only occurs as the result of a very small level of stereoimperfection by the aldolase which can be estimated to be lower by several orders of magnitude (> 10^{-2}–10^{-3}). The fact that the *threo* configurated **79** can be virtually completely transformed into the

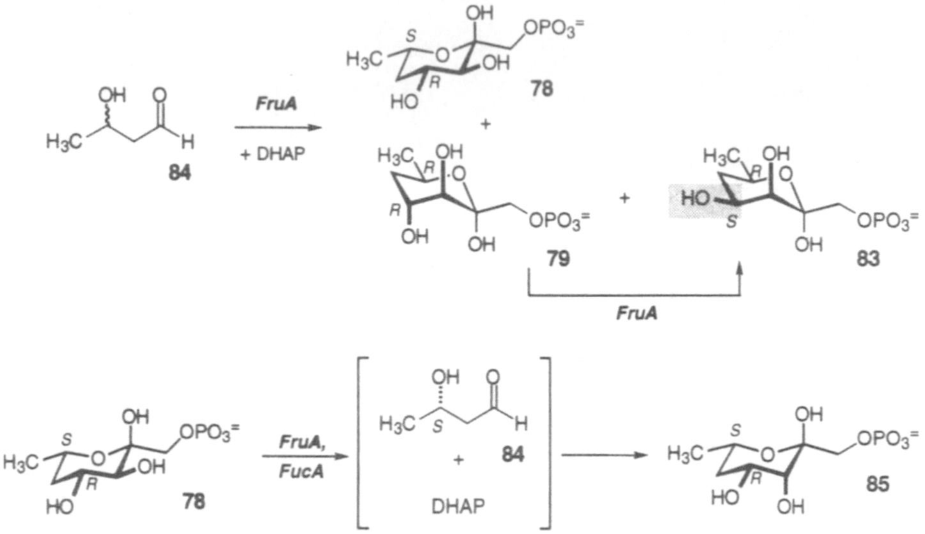

Scheme 18. Stereoinversion at C-4 during FruA-catalyzed aldol additions to 3-hydroxybutanal during long-time equilibration, and stereochemical proof by the formation of an enantiomeric FucA product

inversely *erythro* configurated compound **83** — despite similar thermodynamic stability — testifies that this conversion must be essentially irreversible and that therefore the cleavage reaction for the latter is not catalyzed by the FruA aldolase. This may be rationalized by assuming either that the aldolase is unable to bind *erythro* compounds at all, that the substrate is bound in some reoriented fashion where catalytic activation by crucial residues is inaccessible, or that the reaction path is blocked by energetically highly unfavorable transition states, not readily permissible by the active-site environment for steric reasons (e.g. encumbered boat conformations).

The scope and synthetic usefulness of FruA-catalyzed reactions may be further illustrated by exemplary applications in the preparation of some interesting compounds that comprise functionalities of a different nature. Phosphorylated intermediates of the pentose phosphate pathway, i.e. rare heptulose to nonulose 1-mono- or 1,n-bisphosphates, including analogs of sialic acids and KDO (e.g. **86–88**) (which are difficult to isolate from natural sources for their sensitivity, low abundance, and difficult separability), can be obtained by diastereoselective chain extension of chiral pool carbohydrates or their corresponding phosphates (Scheme 12) [297, 303–305]. Appealing are also strategies for the de novo synthesis of differently substituted, unsaturated (Scheme 16) [296], or regiospecifically labeled [287] sugars. Unusual spiro annulated (**89,90**) and branched chain sugars (**91,92**) have been synthesized [230] from the corresponding readily accessible aldehyde precursors using the multienzymatic in situ preparation of DHAP.

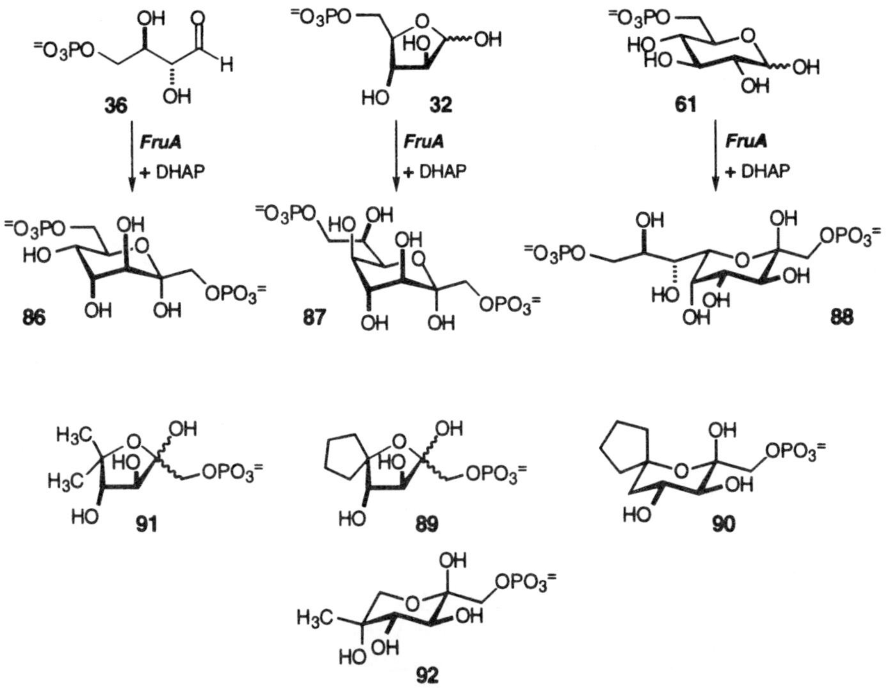

Being restricted to DHAP as the nucleophile, aldol additions will only generate ketoses and derivatives from which aldose isomers may be obtained by biocatalytic ketol isomerization (cf. Sect. 7.1) [306]. For a more direct entry to aldoses the "inversion strategy" may be followed (Scheme 19) [290] which utilizes monoprotected dialdehydes. After aldolization and stereoselective chemical or enzymatic ketone reduction, the remaining masked aldehyde function is deprotected to provide the free aldose. Further examples of the directed, stereodivergent synthesis of sugars and related compounds such as aza- or thiosugars are collected in Sect. 7.

The 3-deoxy D-*arabino*-heptulosonic acid 7-phosphate (DAHP, **34**), intermediate of the shikimic acid pathway (cf. Sect. 2.2.4), has also been prepared

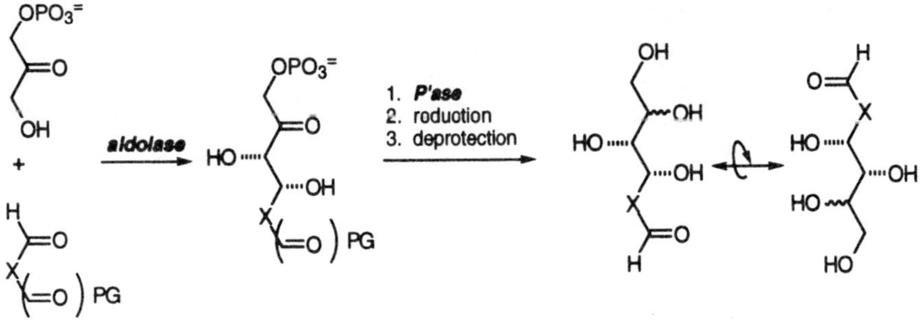

Scheme 19. 'Inversion strategy' for the aldolase-based synthesis of aldoses

based on a FruA-catalyzed aldol reaction [307]. The combined chemical and enzymatic approach employed N-acetylaspartic semialdehyde **93**, derived from racemic C-allyl glycine, as the aldol acceptor. Further elaboration of the resulting adduct **94** produced bioactive **34** and thus proved the correct D- *threo* diol stereochemistry created upon C–C bond formation.

Unusual non-hydrolyzing homonucleoside analogs **95** have been synthesized from a suitably adenine substituted lactaldehyde precursor **96** [308]. Due to its thermodynamic (and possible kinetic) advantage, the D-fructose related diastereomer was formed predominantly (9:1). The L-*sorbo*-configurated isomer **95** resembling adenosine was obtained selectively, however, when the enantiomerically pure (S)-aldehyde served as the substrate.

A sequence of Grignard addition to the 5-chloro-5-deoxy-D-*threo*-pentulose **97**, which is conveniently generated from chloroacetaldehyde **98** by FruA-catalyzed aldolization, and radical cyclization have been designed to provide branched-chain cyclitols **99** [288]. Under modified conditions, the same aldol adduct can alternatively be directed to afford a C-glycosidic furanoside **100**. The latter compound **100** serves as a correctly configured precursor for a Wacker-type manipulation into the spirocyclic *Streptomyces* metabolite sphydrofuran **101** [309]. In an approach resembling the "inversion strategy" an α-C-mannoside **102** has been prepared by a straightforward and stereoselective silane reduction of the unprotected octulose **103**, obtained by DHAP addition to D-ribose 5-phosphate **39** [310].

The key step in the synthesis of the "non-carbohydrate" C9–C16 fragment of pentamycin **104** is the FruA-mediated stereoselective addition of DHAP to an aldehyde precursor, thereby crafting a polyol-type moiety as part of the skipped polyol chain of the macrolide antibiotic (**105, 106**) [311, 312]. The fact that applications of the aldol method are not restricted to carbohydrates or carbohydrate-derived materials — despite obvious advantages — is even more impressively illustrated by the FruA-catalyzed chemoenzymatic syntheses of (+)-*exo*-brevicomin (**107**), the aggregation pheromone of the Western pine bark beetle *Dendroctonus brevicomis* [313]. Addition to DHAP to 5-oxohexanal (**108**) generated the vicinal *syn*-diol structure **109**, comprising the only independent stereogenic centers of brevicomin. Internal acetalization to the bicyclic skeleton followed by deoxygenation of the side chain furnished the optically pure target compound **107**. In an inverse approach [314], the correct stereochemistry present in 5,6-dideoxy-D-*threo*-hexulose **110** has been exploited in its conversion to the identical brevicomin diastereoisomer. The dideoxy *keto* sugar

pentamycin **104**

110, which is readily prepared by DHAP addition to propionaldehyde **111** on the mole scale followed by dephosphorylation [202] or by a complementary route based on transketolase chemistry (Sect. 6.5), was transformed into the pheromone **107** by standard Wittig chain-extension methodology.

5.3 D-Tagatose 1,6-Bisphosphate Aldolase

The D-tagatose 1,6-bisphosphate aldolase is an enzyme that still needs to be classified (TagA; EC 4.1.2.n). In vivo, the Schiff base forming subtype is known to be involved in the catabolism of lactose and D-galactose of different *Staphylococcus* [315, 316], *Streptococcus* [317–319], *Lactococcus* [317], and *Lactobacillus* [320] species as well as in mycobacteria [321]. Zn^{2+}-dependent subtype II TagA enzymes have been found in *Klebsiella pneumoniae* [322] and *Escherichia coli* [323] to be responsible for the ability to grow on galactitol. Enzymes of both class I [324–326] and class II [233, 327, 328] types have been isolated from different sources, and the genes coding for TagA have been cloned from several *Coccus* strains [329–334] and from *E. coli* [335]. Apparently, the class I enzymes from *Staphylococcus* and *Streptococcus* species have no stereochemical selectivity with regard to D-tagatose and D-fructose configurations [324, 326]. Conversely, the class II TagA from *E. coli* has recently been found to be highly stereoselective for its natural substrate in both cleavage and synthesis directions [233, 328].

This has enabled its application in an expeditious multienzymatic synthesis of D-tagatose 1,6-bisphosphate **43**, the (3S,4S) all-*cis* configurated natural substrate, starting from dihydroxyacetone **56** (Scheme 20) [233, 328]. The protocol was based on the ATP-dependent phosphorylation followed by triose phosphate equilibration as discussed above. Concomitant formation of increasing quantities of the unwanted but more stable *fructo*-configurated isomer **42**, arising as a side product owing to an incomplete stereocontrol of the aldolase, necessitated an additional enzymatic step to differentiate the bisphosphates **42/43** in order to facilitate product purification. In situ treatment by yeast fructose-1,6-bisphosphatase (EC 3.1.3.11) or a separate FruA-induced decomposition reaction worked smoothly to specifically convert **42** into monophosphates (**38**) that were readily separated from **43**. When a range of aldehydes were tested as substrate analogs, however, the slight stereochemical infidelity (1%) observed with the natural substrate now became the predominating pathway since not the expected D-*tagato* but rather the thermodynamically more stable D-*fructo* configurated adducts were formed with high preference (≥ 90%). Obviously, the TagA enzyme will have to undergo suitable protein engineering in order to improve its stereoselectivity for broader synthetic use.

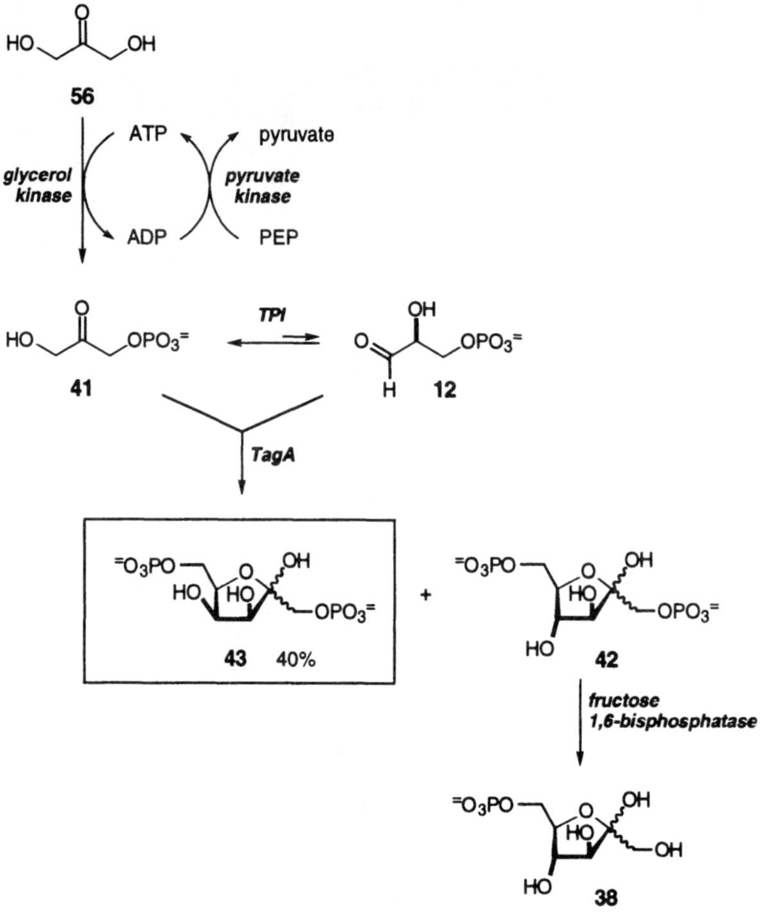

Scheme 20. Multienzymatic synthesis of D-tagatose 1,6-bisphosphate and its differentiation from the *fructo*-configurated side product

5.4 L-Rhamnulose 1-Phosphate Aldolase

The L-rhamnulose 1-phosphate aldolase (RhuA; EC 4.1.2.19) is found in the microbial degradation of L-rhamnose which, after conversion into the corresponding ketose 1-phosphate **44**, is cleaved into **41** and L-lactaldehyde (L-16). The RhuA has been isolated from *E. coli* [336–339], and characterized as a metalloprotein [194, 340, 341]. Cloning was reported for the *E. coli* [342, 343] and *Salmonella typhimurium* [344] genes, and construction of an efficient overexpression system [195, 220] has set the stage for crystallization of the homotetrameric *E. coli* protein for the purposes of an X-ray structure analysis [345].

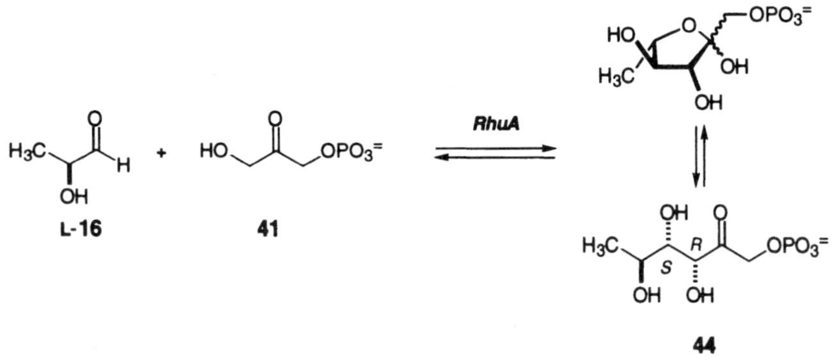

The *E. coli* RhuA enzyme, which has just become commercially available, was surveyed in some detail to evaluate its synthetic potential in the preparation of a range of rare or unnatural sugars having a common (3R, 4S)-*trans* stereochemistry [195, 220]. This study revealed a very high stability in the presence of low Zn^{2+} concentrations with half-lives in the range of months at room temperature, even in the presence of considerable amounts (30–50%) of organic cosolvents such as DMSO, DMF, and ethanol (Fig. 8). The RhuA has a broad substrate tolerance very similar to that of the FruA enzyme with conversion rates generally being usefully high (Table 5). Characteristically, of all DHAP aldolases investigated so far, the RhuA has the greatest tolerance towards sterically congested acceptor substrates, as exemplified in the conversion of the tertiary 2,2,-dimethyl-3-hydroxypropanal [220]. Unbiased kinetic stereocontrol for *re*-face addition to the aldehyde carbonyl is usually observed in RhuA-catalyzed reactions with a variety of 2- or 3-hydroxylated aldehydes (Table 5) [195]. A major difference to the type I FruA enzymes is the concomitant very high kinetic enantiodiscrimination by the RhuA with an overwhelming preference (d.e. $\geq$ 90) for the L-configured antipodes of racemic 2-hydroxyaldehydes D,L-**112** (Fig. 12) [346, 347]. Thus, a number of rare L-ketose 1-phosphates **113** containing 3 contiguous chiral centers are readily accessible from achiral **41** and

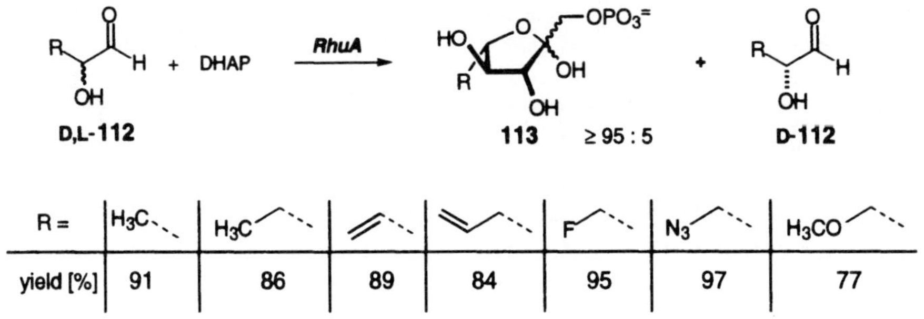

R =	H₃C	H₃C			F	N₃	H₃CO
yield [%]	91	86	89	84	95	97	77

Fig. 12. Kinetic enantioselectivity of the RhuA for L-configured α-hydroxyaldehyde substrates

Table 5. Substrate tolerance of L-rhamnulose 1-phosphate and L-fuculose 1-phosphate aldolases [195, 347]

$$\underset{R}{\overset{O}{\parallel}}\!\!\diagdown\!\!_{H} \; \underset{DHAP}{\overset{aldolase}{\rightleftharpoons}} \; R\diagup\!\!\overset{OH}{\diagup}\!\!\overset{O}{\diagdown}\!\!\diagup OPO_3^= \; + \; R\diagup\!\!\overset{OH}{\diagup}\!\!\overset{O}{\diagdown}\!\!\diagup OPO_3^=$$

	RhuA			FucA		
R	Rel. rate [%]	Selectivity (threo:erythro)	Yield [%]	Rel. rate [%]	Selectivity (threo:erythro)	Yield [%]
L-CH$_2$OH–CH$_3$	100	> 97:3	95	100	< 3:97	83
CH$_2$OH	43	> 97:3	82	38	< 3:97	85
D-CHOH–CH$_2$OH	42	> 97:3	84	28	< 3:97	82
L-CHOH–CH$_2$OH	41	> 97:3	91	17	< 3:97	86
CH$_2$–CH$_2$OH	29	> 97:3	73	11	< 3:97	78
CHOH–CH$_2$OCH$_3$	–	> 97:3	77	–	< 3:97	83
CHOH–CH$_2$N$_3$	–	> 97:3	97	–	< 3:97	80
CHOH–CH$_2$F	–	> 97:3	95	–	< 3:97	86
H	22	–	81	44	–	73
CH$_3$	32	69:31	84	14	5:95	54
CH(CH$_3$)$_2$	22	97:3	88	20	30:70	58

racemic **112** by a convenient one-pot procedure as single diastereomers in high yield [189].

With simple aliphatic aldehydes under conditions of kinetic control the formation of a variable, but considerable fraction (e.g. up to 31% with acetaldehyde) of the unexpected diastereomers epimeric at C-4 was noted [189]. Absolute (3R,4R)-*cis* configuration caused by a reversed, *si*-face attack of the DHAP enolate on the aldehyde carbonyl was proved by comparison with authentic products from respective FucA-catalyzed reactions. Extended studies with substrates covering a diversity of substituents so far revealed no conclusive pattern for substrate control of stereoselectivity [220, 348]. For example, an interesting, virtually complete reversal of stereoselectivity was observed upon terminus extension of the conformationally restricted dialdose acetonides **114** and **115** [348]. From reaction with the vicinal *trans*-hydroxyaldehyde **116** only the expected L-*threo* configurated ulosose derivative **114** was formed, despite the thermodynamically unfavorable diaxial configuration enforced by the bicyclic scaffolding. In contrast, the epimeric vicinal *cis*-hydroxyaldehyde **117** afforded the D-*erythro* diastereomer **115**, as verified by X-ray analysis. However, with a stereochemically related, albeit pyranoid, *galacto*-dialdose derivative **118**, which was produced in situ by the action of galactose oxidase (GalO; EC 1.1.3.9) [349], stereoregular *threo*-connection to **119** resulted [350]. Clearly, an extended basis of aldehydic probes will be necessary — ideally together with an X-ray structure of the enzyme — to better define stereochemical requirements for matched or mismatched cases.

In contrast to the FruA preference for phosphorylated substrates, anionically charged aldehydes are not converted by the RhuA [220]. For example, glyceraldehyde phosphates of both D- or L-configurations are not accepted, nor are other sugar phosphates. Similarly, glyoxylic acid is not a substrate, although the less acidic succinic semialdehyde reacts [200]. Adducts formally derived from glyoxylic acid may be obtained in good yield, however, by submission of the corresponding alkyl esters under buffered reaction conditions, since the ester function becomes hydrolyzed during (or after) C–C bond formation and the carboxylic acid so produced defies cleavage by the aldolase [208].

The scheme of thermodynamic equilibration, particularly for 3-hydroxyaldehydes 75, as described above for FruA catalysis (Scheme 16) can likewise be applied to reactions with RhuA, the "enantiomeric" enzyme [189, 220]. As is to be expected from the complementarity of the stereoselectivities of these enzymes, the corresponding enantiomeric compounds ent-78 are enriched upon equilibration. Owing to the lower diastereoselectivity of the RhuA versus the FruA enzyme, however, the corresponding FucA configured isomer 85 accumulates more rapidly at the expense of the thermodynamically less favored diastereomer ent-79 (cf. Scheme 18, Sect. 5.2).

5.5 L-Fuculose 1-Phosphate Aldolase

Like its RhuA companion, the L-fuculose 1-phosphate aldolase (FucA; EC 4.1.2.17) functions in a closely parallel metabolic sequence for microbial degradation of L-fucose via the corresponding ketose 1-phosphate **45** which results in the same DHAP (**41**) and L-lactaldehyde (L-**16**) fragments. Early enzymologic work was reported for the FucA purified from an *E. coli* strain grown on L-fucose [351, 352]. Its structural gene has been cloned [353, 354] and utilized for highly efficient overexpression independently by two groups [195, 355]. The protein, which is active as a homotetramer, has been crystallized, and the 3D structure has been determined by X-ray analysis at 2.13 Å resolution [211]. The subunits, which in a tetramer are arranged in C_4-symmetry (Fig. 13), are built up from a nine-stranded β-pleated sheet which is covered on both sides by two and three α-helices (Fig. 14). The active site is located at the subunit interface and opens to the outside of the tetramer. The catalytic zinc ion is tightly coordinated by three N^ε-atoms of histidine residues (His92, His94, and His155) and by a bidentate contact from a glutamic acid residue (Glu73).

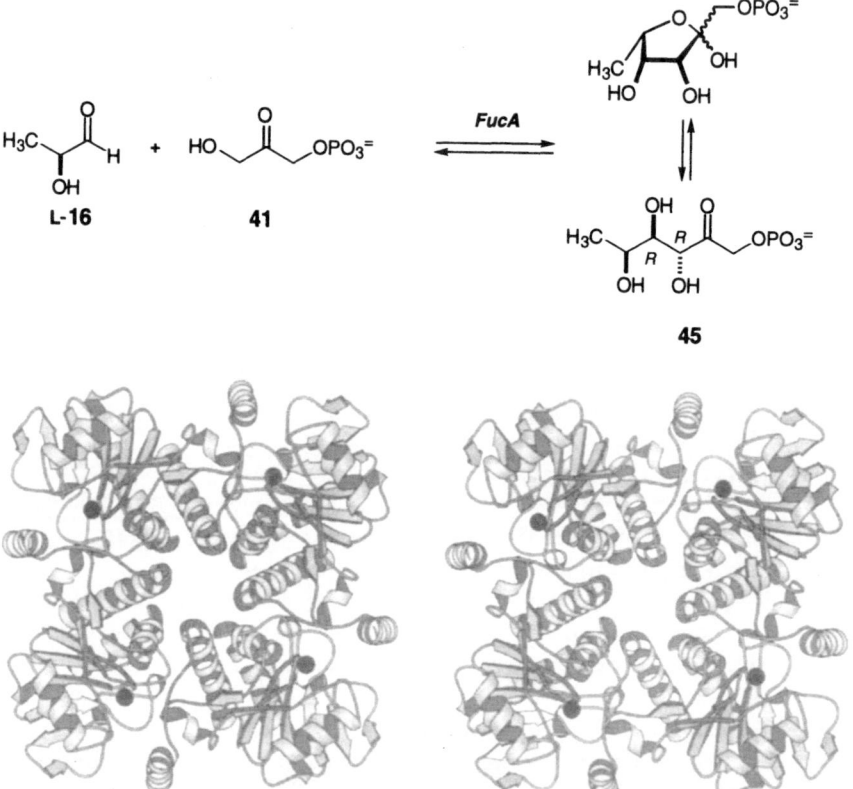

Fig. 13. Stereo ribbon plot [76] of the tetrameric FucA from *Escherichia coli* along its molecular 4-fold axis. The location of the catalytic zinc ions is marked by black circles

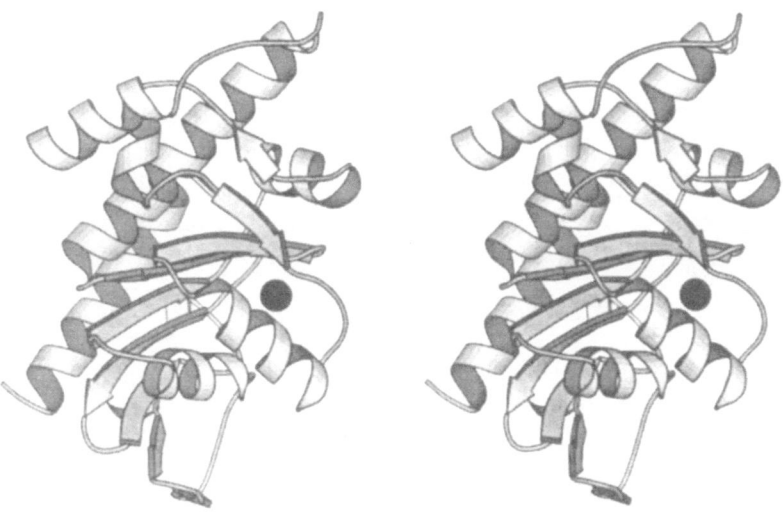

Fig. 14. Stereo ribbon representation [76] of the FucA subunit architecture showing the central nine-stranded β-pleated sheet. The active site zinc ion is marked by a black circle

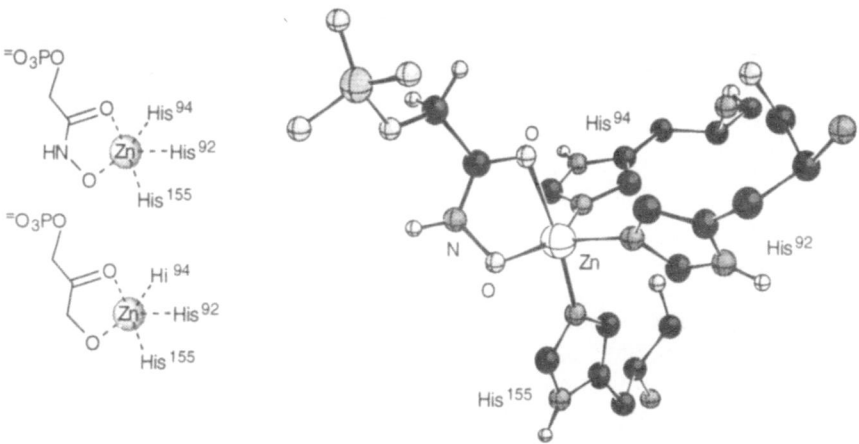

Fig. 15. Coordination sphere of the zinc ion in the FucA crystal structure liganded by the chelating inhibitor phosphoglycolohydroxamate, which is a transition state analog for activated DHAP

More recently, a corresponding structure has been determined for the enzyme containing the substrate analog phosphoglycolohydroxamate (Fig. 15) [356], which is an inhibitor of DHAP-dependent metalloaldolases in the nM range [357]. The latter analysis revealed that the inhibitor is bound as a replacement of Glu73 in a chelating fashion involving both hydroxamate oxygens, and thus clearly points to a bidentate binding of DHAP at the active-site zinc ion in an enediolate state in order to provide the activation required for the C–C bond-forming step [357] (Fig. 16). This result stands in sharp contrast to

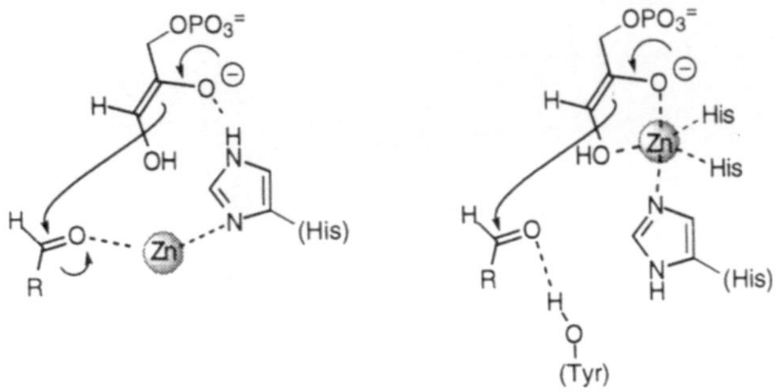

Fig. 16. Mechanistic model for DHAP addition to an aldehyde acceptor by class II aldolases based on the inhibitor-liganded FucA structure (right) in comparison to earlier proposals (left)

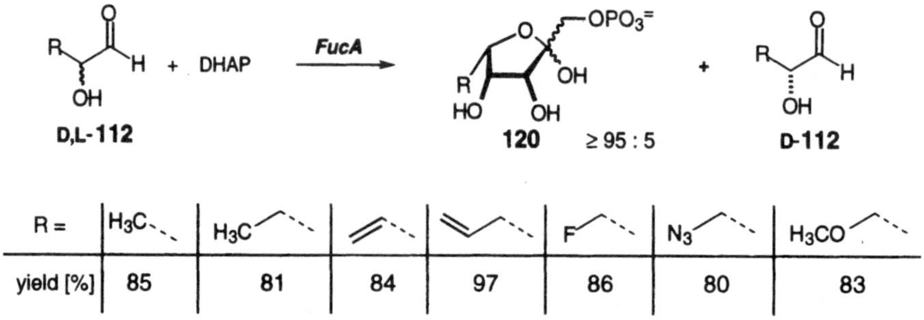

R =	H₃C	H₃C			F	N₃	H₃CO
yield [%]	85	81	84	97	86	80	83

Fig. 17. Kinetic enantioselectivity of the FucA for L-configurated α-hydroxyaldehyde substrates

accepted earlier mechanistic models for type II aldolases that had been built upon investigations on yeast aldolase (FruA). Based upon ESR and NMR relaxation rate measurements on the Mn^{2+}-substituted holoenzyme, it has been suggested that the binding of DHAP occurs through its phosphate group [358], and carbonyl polarization by the metal ion via an intervening imidazole relay [359, 360]. Subsequent FT-IR [361] and deuterium exchange rate studies [47] with native yeast aldolase had concluded that there was a possible extra direct carbonyl coordination for the aldehyde activation which, if structural features of the FucA were of general implication for type II aldolases, probably rather occurs through protonation by acidic residues.

Like a number of other aldolases the FucA enzyme is now also offered commercially. Overall practical features make the FucA quite similar to the RhuA enzyme, as is illustrated by its high stability in the presence of Zn^{2+} ions, by its broad substrate tolerance for variously substituted aldehydes at useful reaction rates (Table 5), and by a high asymmetric induction for (3R,4R)-cis stereoselectivity by si-face addition to the aldehyde carbonyl [195, 355]. Although incomplete stereochemical fidelity is also observed in reactions with certain, mostly aliphatic, aldehydes [195] the diastereomeric ratios are usually higher, and diastereospecific results observed more often, with the FucA than with the RhuA enzyme [362]. Adding to its synthetic value, the FucA offers powerful kinetic resolution of 2-hydroxylated substrates **rac-112** for a simultaneous determination of three contiguous chiral centers [346, 347]. Again, this allows a one-pot formation of diastereochemically pure L-configurated adducts **120** (d.e. $\geqslant 90$) from racemic starting materials (Fig. 17), e.g. in the preparation of a number of rare or unnatural sugars and deoxyazasugars (Sect. 7).

6 Miscellaneous Lyases

Of the range of lyases known, but not included in the sections above for their distinct mode of action, an increasing number are being evaluated for their potential applicability to preparative synthesis. Since earlier restrictions of limited catalyst accessibility or inconvenience of cofactor requirements are being removed at a rapid pace by recent genetic advancements, other selected lyases have been studied in detail. The most promising examples and their applications to asymmetric synthesis have been included in this review and will be discussed below.

6.1 2-Deoxy-D-ribose 5-Phosphate Aldolase

Functionally and mechanistically reminiscent of the pyruvate lyases, the 2-deoxy-D-ribose 5-phosphate (**121**) aldolase (RibA; EC 4.1.2.4) [363] is involved in the deoxynucleotide metabolism where it catalyzes the addition of acetaldehyde (**122**) to D-glyceraldehyde 3-phosphate (**12**) via the transient formation of a lysine Schiff base intermediate (class I). Hence, it is a unique aldolase in that it uses two aldehydic substrates both as the aldol donor and acceptor components. RibA enzymes from several microbial and animal sources have been purified [363–365], and those from *Lactobacillus plantarum* and *E. coli* could be induced to crystallization [365–367]. In addition, the *E. coli* RibA has been cloned [368] and overexpressed. It has a usefully high specific activity [369] of 58 U mg^{-1} and high affinity for acetaldehyde as the natural aldol donor component ($K_m = 1.7$ mM) [370]. The equilibrium constant for the formation of **121** of 2×10^{-4} M does not strongly favor synthesis. Interestingly, the enzyme's relaxed acceptor specificity allows for substitution of both cosubstrates: propionaldehyde **111**, acetone **123**, or fluoroacetone **124** can replace **122** as the donor [370, 371], and a number of aldehydes up to a chain length of 4 non-hydrogen atoms are tolerated as the acceptor moiety (Table 6).

2-Hydroxyaldehydes are relatively good acceptors, and the D-isomers are superior to the L-isomers [370]. It is noteworthy that from the reaction with **111** as the donor only a single diastereomeric product (e.g. **126**) of absolute (2R,3S) configuration results [370], indicative not only of the high level of asymmetric induction at the newly formed stereogenic center at C-3, but also of a stereospecific deprotonation at C-2 of the donor.

A salient shortcoming of the catalyst is the relatively low conversion rate obtained with any of the substrate analogs tested (< 1%) which in practice must be compensated by utilization of large amounts of enzyme and long reaction times. Thus, in reactions leading to thermodynamically unfavorable products it has been observed under equilibrating conditions that — in common with results discussed above for other aldolases — additions do not proceed fully stereospecifically at the reaction center [369].

It has been realized serendipitously that the initial product from aldolization of **122** to itself again serves as a suitable acceptor for sequential addition of a second donor molecule to give (3R,5R)-2,4,6-trideoxyhexose **127** in a 20% yield [372]. By variation of the first acceptor in the form of substituted acetal-

Table 6. Substrate tolerance of deoxy-D-ribose 5-phosphate aldolase

R	Yield [%]	Rel. rate [%]	Ref.
$CH_2OPO_3^=$	78	100	[370]
H	20	–	[372]
CH_2OH	65[a]	0.4	[370]
CH_3	32	0.4	[370]
CH_2F	33	0.4	[370]
CH_2Cl	37	0.3	[370]
CH_2Br	30	–	[373]
CH_2SH	33[a]	–	[373]
CH_2N_3	76	0.3	[370]
C_2H_5	18	0.3	[370]
$CH=CH_2$	12	–	[373]
$CHOH-CH_2OH$	62[a]	0.3	[370]
CHN_3-CH_2OH	46	–	[373]
$CHOH-CH_3$	51[a]	–	[373]
$CHOH-CH_2-C_6H_5$	46[a]	–	[373]
$CH_2SCH_2-CHOH-CH_2OH$	27	–	[373]

[a] Equilibrates to a more stable pyranose isomer.

dehydes, several related double aldol adducts could be prepared. Formation of higher order adducts is effectively precluded by rearrangement into stable cyclic hemiacetals and thus masking of the requisite free aldehyde forms.

6.2 3-Hexulose 6-Phosphate Aldolase

Like the DHAP aldolases, the class II 3-hexulose 6-phosphate aldolase from a unique formaldehyde-fixing system of the methylotrophic bacterium *Methylomonas* M15 utilizes a ketose phosphate, i.e. D-ribulose 5-phosphate (**128**), as the aldol donor component for which it has a stringent requirement [374]. On the other hand the enzyme accepts a wide variety of aldehydes (some 20 examples have assayed positive) as a replacement for formaldehyde. The latter is considered to be the natural acceptor employed in the ribulose monophosphate metabolic cycle [375].

For preparative applications, the expensive and configurationally unstable donor 128 can be simply prepared in situ by the action of ribose 5-phosphate isomerase (EC 5.3.1.6) on D-ribose 5-phosphate (39). This technique was applied to the stereoselective synthesis of D-[1-^{13}C] fructose 6-phosphate 38 from [^{13}C] formaldehyde [376, 377] which also included a second enzymatic isomerization of the D-arabino-3-hexulose 6-phosphate intermediate 129 into the more stable 2-hexulose derivative 38. Notable are the conflicting demands for high substrate levels (necessary to shift the fully reversible multi-component equilibrium) versus the notorious enzyme inactivation that occurs at higher formaldehyde concentrations.

Addition to formaldehyde [378] and other aldehydes [379, 380] proceeds with high absolute stereoselectivity for the (5R)-configurated products (configurational reference changes for any substrate larger than formaldehyde!). In contrast, no or only a low level of relative acceptor diastereoselectivity at the chiral C-6 was determined in the reactions with acetaldehyde (6S/6R ~ 1:1) [379] and propionaldehyde (130/131 = 1:2.4) [380] as the stereochemical probes.

6.3 Dimethylmalic Acid Lyase

The degradation of nicotinic acid by *Clostridium barkeri* involves the cleavage of the intermediate 2,3-dimethylmalate **132** from which propionic and pyruvic acids are formed by a specific lyase (EC 4.1.3.32). In the reverse direction, the enzyme must have the unusual capacity to deprotonate propionic acid at the α-carbon instead of the carboxylic acid function, or next to an anionic carboxylate. Purified dimethylmalic acid aldolase has been used to catalyze the stereospecific addition of **133** to the oxoacid acceptor, yielding the (2*R*,3*S*) configurated dimethylmalic acid **132** at the multi-gram scale [381]. The substrate tolerance of this enzyme has not yet been determined.

6.4 Transaldolase

The transaldolase (EC 2.2.1.2) is an ubiquitous enzyme that is involved in the pentose phosphate pathway of carbohydrate metabolism. The class I lyase, which has been cloned from human [382] and microbial sources [383], transfers a dihydroxyacetone unit between several phosphorylated metabolites. Although yeast transaldolase is commercially available and several unphosphorylated aldehydes have been shown to be able to replace the acceptor component, preparative utilization has mostly been limited to microscale studies [384, 385] because of the high enzyme costs and because of the fact that the equilibria usually are close to unity. Also, the stereochemistry of transaldolase products (e.g. **38**, **40**) [386] matches that of the products from the FruA-type DHAP aldolase which are more effortlessly obtained.

Research into the enzymatic conversion of starch **134** into D-fructose **37** [387] was hampered by the inability of the authors to find a phosphatase that was specific for fructose 6-phosphate **38**. The task could be completed by constructing a salvage cycle in which transaldolase transferred the dihydroxyacetone moiety from **38** onto D-glyceraldehyde **135** with the release of D-fructose **37**. The byproduct D-glyceraldehyde 3-phosphate **12** was finally dephosphorlyated using a specific phosphoglycerate phosphatase.

Transaldolase catalysis has been used for the formation of D-*threo*-2,5-diketohexose **136** from fructose 6-phosphate **38** and hydroxypyruvaldehyde **137** [388]. Because of the accumulation of the by product **12**, at equilibrium a conversion of merely 50% of the starting materials was reached. By reducing the concentration of **12** through coupled isomerization to DHAP and subsequent aldol addition to the stable **42**, product flux could be strongly enhanced to a level at which starting sugar phosphate was no longer detectable [348]. Effective utilization of **137** required the use of the FruA from *E. coli* (instead of the rabbit muscle enzyme) because of its very low competing acceptance of nonionic aldehydes.

6.5 Transketolase

A number of "lyases" are known which, unlike the aldolases, require thiamine diphosphate (TDP) as a cofactor in the transfer of acyl anion equivalents [389–391], but proceed via enolate-type intermediates by a mechanism that resembles the classical benzoin addition. The most important representative is the transketolase (EC 2.2.1.1) [392] which stems from the oxidative pentose phosphate pathway where it catalyzes the reversible transfer of a hydroxyacetyl fragment from a C_5–C_7 ketose phosphate (e.g. **40**, **138**) to a C_3–C_5 aldehyde phosphate (e.g. **12**, **39**). The ubiquitous enzyme can be readily isolated either from spinach leaves [393], from baker's yeast (which can be crystallized [394] and which is commercially available), from recombinant *E. coli* strains [395, 396], or from mammalian liver [397]. The transketolase gene has been cloned from prokaryotes (*E. coli* [398, 399], *Rhodobacter sphaeroides* [400]), from yeasts (*Saccharomyces cerevisiae* [401, 402], *Pichia stipitis* [403]), and from human liver [404]. The crystal structure of the *S. cerevisiae* transketolase has been determined at 2.0 Å resolution [405, 406], and the function of the TDP cofactor has been further elucidated by solving structures of the apoenzyme complexed with TDP analogs including one that resembles a reaction intermediate [407, 408]. The enzyme is a dimer (Fig. 18) with the active sites located at the interface of the subunits, each made up from 680 amino acid residues. The Ca^{2+}-complexed cofactor TDP is completely shielded from solvent access except for the catalytically active thiazolium C-2 atom (Fig. 19).

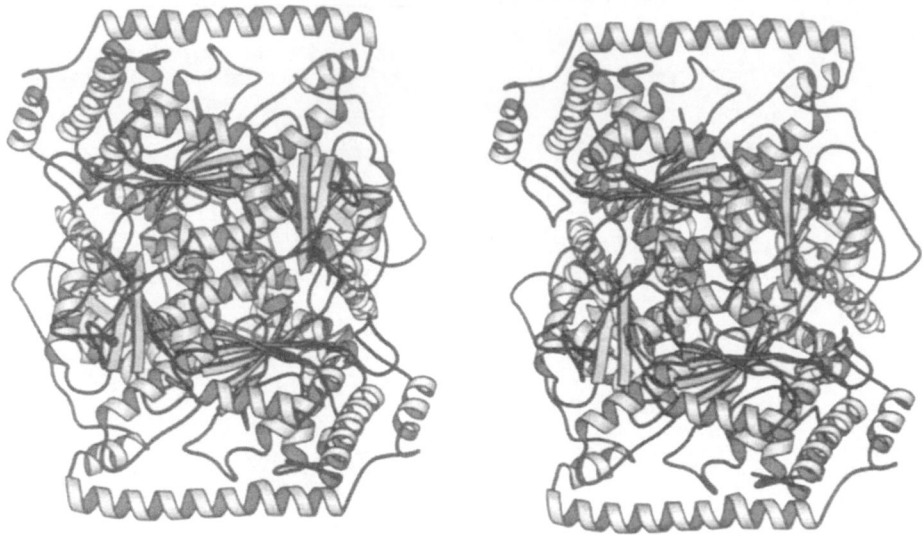

Fig. 18. Stereo ribbon representation [76] of the transketolase dimer from *Saccaromyces cerevisiae*

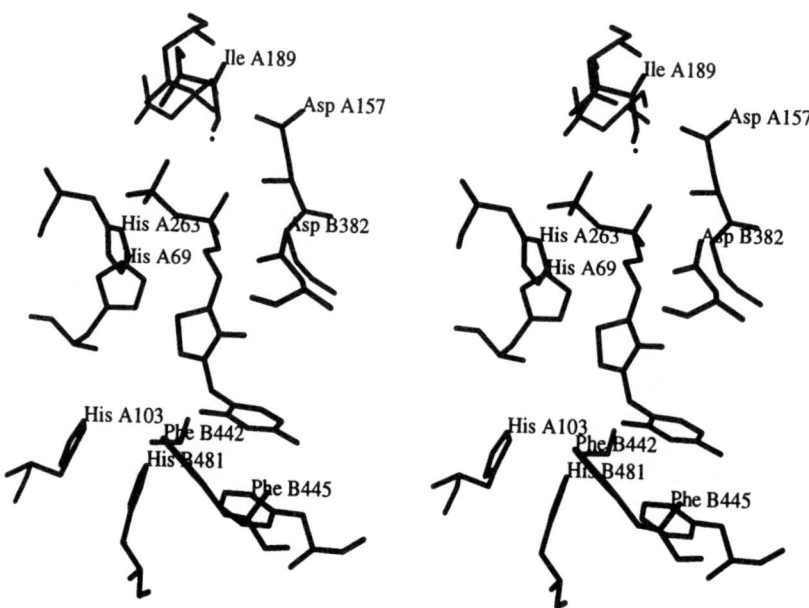

Fig. 19. Stereo view of the active site of the transketolase from *Saccaromyces cerevisiae* showing the environment of the thiamine diphosphate cofactor. The magnesium ion bridging the diphosphate unit is represented as a dot

Because of its ready availability and stability, the transketolase from yeast has been studied more frequently than enzymes from other sources for mechanistic as well as synthetic purposes. A valuable feature is that during the first round of a *ping-pong* catalytic cycle the natural ketose donor component(s) can be effectively replaced by hydroxypyruvate (139) [394]. In this process, the reactive intermediate is formed by a spontaneous decarboxylation that renders the overall preparative addition to aldehydic substrate analogs essentially irreversible. On a preparative scale, this usually compensates for a reduced v_{max} (e.g. about 25 times less as compared to 138 [409]) since 139 seems not inhibitory even at higher concentrations [409]. Although commercial 139 is quite expensive, a multi-gram synthesis from pyruvate has been detailed [410,411] which can be readily scaled up to provide mole quantities [412].

The studies revealed a rather broad acceptor tolerance in that the natural acceptor aldehyde(s) can be considerably varied among phosphorylated as well as unphosphorylated α-hydroxyaldehydes 140 which are both converted at comparable rates (Table 7).

Transketolase adds the hydroxyacetyl moiety stereospecifically to the re-face of the acceptor leading to products of (3S) configuration. Although the catalytic reaction creates only a single stereocenter, the enzymes from yeast, spinach, or E. coli can efficiently distinguish between the antipodes of rac-140 with a seemingly unbiased preference for a (2R)-hydroxyl group, leaving enantiomerically pure (S)-configured α-hydroxyaldehydes (S)-140 behind [416,417]. In effect, the high kinetic enantioselectivity makes it possible to control the stereochemistry of two adjacent stereogenic centers in the generation of enantiomerically and diastereomerically homogenous (3S,4R)-configured ketose-type products 141 when starting from racemic aldehydes. This two-carbon elongation method thus provides the D-threo products 141, equivalent to those created by the respective three-carbon elongation using FruA catalysis (cf. Sect. 5.2). A study on the stereoselectivity of spinach transketolase with achiral substrates such as generic or CH$_3$O- and CH$_3$S-substituted aliphatic aldehydes noted reasonable relative reaction rates (e.g. ethanal 12%, methoxyethanal 32%) but the products had an enantiomeric excess of only 60–76% [413]. It appears, however, that the study was flawed by inappropriately buffered reaction and work-up conditions,

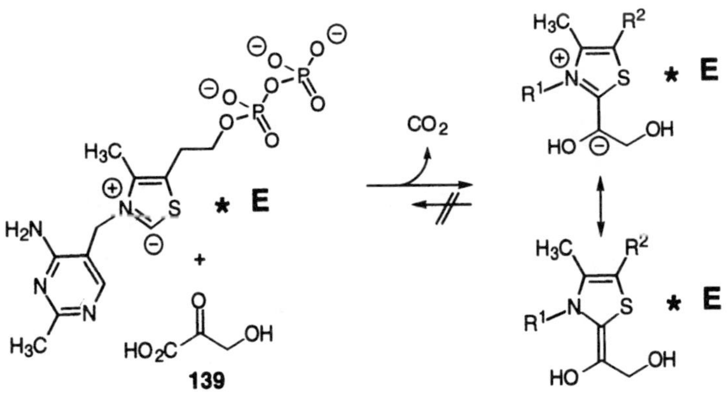

Scheme 21. Thiamine-diphosphate-dependent decarboxylation of hydroxypyruvate by transketolase

Table 7. Substrate tolerance of transketolase

R	Rate [%]	Yield [%]	Ref.
H	100	70	[314, 409, 413]
CH_2OH	37	57[a]	[413]
$CH_2OPO_3^{=}$	–	96	[362, 414, 415]
CH_2OCH_3	27	76	[416, 417]
$CH_2OCH_2C_6H_5$	–	79	[416]
CH_2F	47	79	[416, 418]
CH_2N_3	–	71	[419]
CH_2CN	–	82	[416]
CH_2SH	–	78[a]	[416, 420]
CH_2SCH_3	33	–	[417]
$CH_2SCH_2CH_3$	–	74	[416]
CH_2CH_2OH	< 10	14[a]	[417]
$CHOH–CH_3$	35	50	[421]
CH_3	44	88	[409, 416–418]
CH_2CH_3	33	90	[416, 417]
$CH_2CH_2CH_3$	22	78	[417]
$C(CH_3)_3$	11	–	[417]
$CH=CH_2$	56	60	[417]
$CH_2CH=CH_2$	28	90	[417]
$(S)-CHOH–CH=CH_2$	36	60[a]	[417]
$(R)-CHOH–CH=CH_2$	32	63[a]	[417]
D-ribo-$CHOH–CHOH–CH_2OH$	30	70[a]	[418, 422]

[a] Equilibrates to a more stable furanose or pyranose isomer.

which probably caused partial racemization. Although aromatic aldehydes have been claimed to be substrates of transketolase due to the disappearance of hydroxypyruvate in enzymatic assays, proof for stereoselective product formation is missing so far.

Mostly the baker's yeast transketolase has been used so far to prepare several valuable ketose sugars and derivatives **141** (cf. Sect. 7). Recently, the transketolase was utilized in the key stereogenic transformation of racemic 2-hydroxybutyraldehyde **142** into the homochiral synthon 5,6-dideoxy-D-*threo*-hexulose **110** for the chemoenzymatic synthesis of (+)-*exo*-brevicomin **107** [314]. Transketolase has also been applied for the in-situ generation of Ery4P (**35**) from Fru6P (**38**) in a multi-enzymatic synthesis of **34** (Scheme 5).

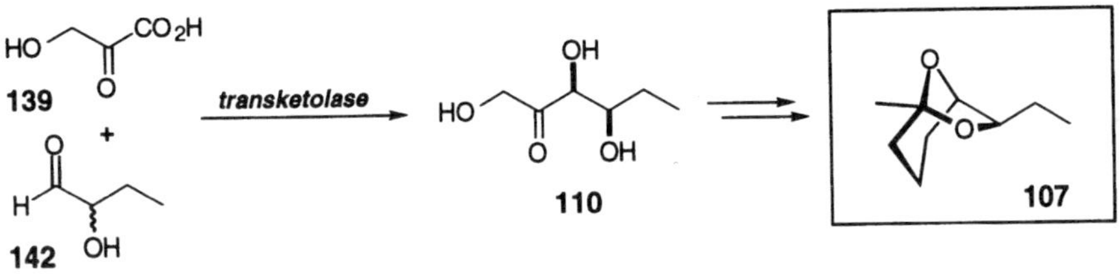

6.6 Pyruvate Decarboxylase

The Baker's-yeast-mediated formation of acyloins has long been exploited in synthesis [423, 424]. It was a long time before the enzymic catalyst responsible for this ability was identified as the thiamine-diphosphate-(TDP)-dependent pyruvate decarboxylase (EC 4.1.1.1) [425, 426], which in vivo decomposes pyruvate into acetaldehyde and carbon dioxide (Scheme 22). The brewer's yeast PDC, which is commercially available, has extensively served as a model for mechanistic investigations. The enzyme can be readily purified from yeasts [427, 428], wheat germ [429], pea seeds [430], or from *Zymomonas mobilis* [426, 431] where the PDC is particularly abundant. PDC structural genes have been cloned from several organisms including yeasts [432–434], bacteria [435–439], and plants [440], and an X-ray structure for the enzyme from *Saccharomyces uvarum* has recently been determined at 2.4 Å resolution [441]. The homotetrameric protein (dimer of dimers) is assembled from subunits containing 1126 amino acids. The catalytic centers, which are deeply buried in the subunit interface, contain both TDP and Mg^{2+} cofactors in a binding fold that resembles that of transketolase (Fig. 19) [442].

The intermediary cofactor bound acetyl anion equivalent can be transferred to an aldehyde acceptor, e.g. to acetaldehyde already produced during regular catalytic reaction in which optically active 3-hydroxy-2-butanone (acetoin, an important aroma constituent) is formed. Interestingly, PDCs from different sources differ in stereoselectivity [443]: acetoin (R)-**143** is obtained using brewer's yeast PDC (ee 28–54%) [444, 445] while the enantiomeric (S)-**143** is produced preferentially by PDC from wheat germ (ee 16–34%) [446] or from *Z. mobilis* (ee 24–29%) [445]. When glyoxylate **14** (instead of **2**) is subjected to decarboxylation in the presence of acetaldehyde, optically active lactaldehyde

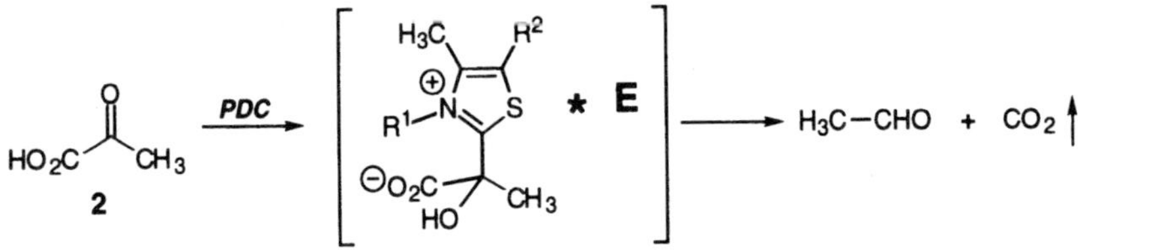

Scheme 22. Thiamine-diphosphate-dependent fragmentation of pyruvate by pyruvate decarboxylase

16 is formed, although the asymmetric induction is modest: ee 19–22% for the (R)-enantiomer obtained with the yeast and ee 52–61% for the (S)-enantiomer obtained with the Z. mobilis enzymes, respectively [445].

With benzaldehyde **144** or halogenated derivatives (Cl, F) as acceptors the yeast-PDC-catalyzed addition proceeds with almost complete stereoselectivity to furnish the corresponding (R)-configured 1-hydroxy-1-phenylpropanones **145** [447]. For practical reasons, whole yeast cells are most often used as the catalyst, with only small loss of enantioselectivity [423, 424]. The conversion of benzaldehyde in particular has gained industrial importance because the acyloin is an important precursor for the synthesis of L-(–)-ephedrine [448]. Otherwise, the substrate tolerance is remarkably broad for aromatic aldehydes; on the laboratory scale, however, yields of acyloins are usually low because of the prior or consequent reductive metabolism of aldehyde substrate and product, giving rise to considerable quantities of alcohol **146** and vicinol diols **147**, respectively [423, 424, 449]. The range of structural variability covers both higher α-oxo-acids (e.g. -butyrate, -valerate) as the donor component, as well as α,β-unsaturated aldehydes (e.g. cinnamaldehyde **148**) as the acceptor [450].

L-ephedrine

6.7. Threonine Aldolase

A subclass of lyases, involved in β-hydroxy-α-amino acid metabolism, utilizes pyridoxal 5′-phosphate (PLP) as a prosthetic group for imine/enamine type activation. These enzymes are not only of interest for the synthesis of the naturally occurring prototypes L-serine (Sect. 6.8) or L-threonine, but also offer a potential entry to rare or non-natural analogs.

In various mammalian tissues an enzymatic activity has been reported [451–453] which causes the liberation of glycine **149** and acetaldehyde from L-threonine **150** and has therefore been named threonine aldolase (ThrA; EC 4.1.2.5). It is curious that *allo*-threonine **151** seems to be a more active substrate for this enzyme than is **150**. Cleavage of L-3-phenylserine is also catalyzed by mammalian ThrA enzymes [454–456], which in direction of synthesis nonselectively produce both *threo* (**152**) and *erythro* (**153**) configurated adducts from benzaldehyde and **149** [454]. The same enzyme preparations were also able to act on **150**. Thus, considerable disagreement still exists in the literature about the true nature of these enzymatic activities.

A corresponding ThrA has been detected in a number of strictly anaerobic bacteria, and the enzyme from *Clostridium pasteurianum* has been purified and shown to be highly selective for L-threonine **150** [457]. A corresponding L-specific catalyst has also been purified and crystallized from cells of the yeast *Candida humicola*. Very recently, the latter enzyme was reinvestigated for synthetic purposes and found to have a very broad substrate tolerance for the aldehyde acceptor, notably including variously substituted aliphatic and aromatic aldehydes [458]. Like other aldolases known up to now, α,β-unsaturated aldehydes were not acceptable. In common with the mammalian enzymes, it was found that the enzyme in fact prefers L-*erythro* to L-*threo* stereochemistry in both the synthesis and cleavage directions. However, the D-configurated compounds were inactive. Diastereoselectivity for (2S,3S) vs. (2S,3R) configuration varied from 93:7 (ethanal), 53:47 (3-benzyloxypropanal, **154**), 40:60 (benzaldehyde) to 30:70 (4-hydroxybenzaldehyde, **155**). For preparative reactions, a large excess of **149** was applied to compensate for the unfavorable equilibrium constant.

A ThrA enzyme that is highly specific for L-*threo* stereoisomers has been detected in *Streptomyces amakusaensis*. The purified enzyme has been employed for the efficient resolution of chemically generated racemic *threo*-phenylserines **156**, substituted at the *p*-position (H, OH, Br, NO$_2$), to furnish enantiomerically pure D-amino acids **157** as the (2R,3S) diastereomers [459]. Preliminary screening results suggest a broad substrate tolerance [460].

6.8. Serine Hydroxymethyltransferase

The pyridoxal-5′-phosphate dependent serine hydroxymethyltransferase (SHMT; EC 2.1.2.1) in vivo catalyzes the interconversion of L-serine **158** and glycine **149** by transfer of the β-carbon of L-serine to tetrahydrofolate (THF) by which the activated formaldehyde is physiologically made available as a C$_1$-pool. The reaction is fully reversible and provides a means for the stereoselective synthesis of **158** in vitro from donor **149** and formaldehyde. Economical yields (88–94%) of L-serine have thus been obtained on a multimolar scale using raw cell extracts of recombinant *Klebsiella aerogenes* or *E. coli* in a controlled bioreactor at final product concentrations $\geqslant 450 \, \mathrm{g\,l}^{-1}$ [461, 462]. Several SHMTs have been purified and characterized from various organisms including animal tissues [463, 464], eucaryotic [465] and procaryotic microorganisms [466–468]. Genes coding for SHMT have been cloned from a variety of mammalian [469–474] and microbial organisms [475–482].

The SHMT enzymes isolated from corn and from rabbit liver [483], as well as that from pig liver [484], were studied for their preparative utility (Table 8). In the absence of THF all enzymes will accept a range of aldehydes to furnish the corresponding β-hydroxy-α-amino acid adducts. An excess of the least costly component (usually glycine) is advisable in order to shift the equilibria towards synthesis [484]. The enzymes from either source showed a high level of asymmetric induction for L-amino acids. The relative diastereoselectivity in creating the hydroxyl bearing chiral center, however, was much less pronounced, since the initial kinetic preference in favor of L-*erythro* isomers **159** was quite rapidly offset by equilibration to give equally or more stable *threo* isomers **160**. This is particularly problematic since preparative scale experiments with substrate analogs required long incubation times to achieve practical yields [484].

6.9 Polycarboxylic Acid Lyases

A number of biosynthetically important enzymes promote aldol related Claisen additions of acetyl thioesters of coenzyme A (CoA) to ketones, often to α-oxoacids. This includes the citrate synthase (EC 4.1.3.7) which is one of the key enzymes in central metabolism, and several others involved in the biosynthesis

Table 8. Substrate tolerance of serine hydroxymethyltransferase

R	Diastereoselectivity (erythro:threo)	Yield [%]	Ref.
H	–	94	[461, 462]
CH$_3$	98:2	–	[485]
(CH$_2$)$_5$CH$_3$	40:60	25	[484]
cyclohex-3-enyl	n.d.	11	[484]
C$_6$H$_5$	40:60	22	[484]
2-furanyl	50:50	20	[484]
2-thienyl	44:56	11	[484]
2-imidazolyl[a]	67:33	10	[484]
2-imidazolyl[b]	36:64	55	[484]

[a] 1 day incubation
[b] 60 days incubation

of fatty acids, steroids, terpenoids, macrolides and other secondary metabolites. Members of this class of C–C bond-forming enzymes would be attractive for asymmetric synthesis, and would offer the added advantage that equilibrium constants reflect the high driving force from the ensuing hydrolysis of the CoA thioester. The high costs and limited stability of the cofactor have so far precluded broader synthetic evaluations, but the technique may be more practical now since several cost efficient in situ cofactor recycling schemes for acetyl-CoA have been developed that are based on either the acetyl donors acetyl phosphate and acetyl carnitine **161** [486, 487] or on acetanhydride in a phase-transfer catalytic system [488]. Using these principles, the synthesis of citric acid **162** from oxalacetate **163** has been demonstrated at the multi-mmol level and this may have practical value for the preparation of specifically labelled material. Some short chain acetyl CoA analogs can be prepared enzymatically by using acetyl-CoA synthetase (EC 6.2.1.1) [489, 490], and another chemoenzymatic approach to CoA analogs based on CoA biosynthetic enzymes was recently described [491].

Advances are currently being made in the area of polyketides, a class of structurally diverse natural products possessing a broad range of biological activities. The biosynthesis involves ketosynthase and functional group processing enzymes that iteratively build up a polyketide backbone. By choosing from the number of known genes that code for the individual enzymes from different microorganisms and plants, new interspecies genetic constructs can thus be assembled in a combinatorial fashion; in this way the feasibility of an engineered biosynthesis of novel, non-natural polyketides with potentially useful bioactivity can be explored [492–500].

6.10. Hydroxynitrile Lyase

Optically active cyanohydrins widely occur in an O-glycosylated form in plants and in many insects where they serve as antifeedants. The formation of cyanohydrins is catalyzed by hydroxynitrile lyases (oxynitrilase) [501] which, according to their stereoselectivity, are divided into the three subtypes (R)-oxynitrilase (EC 4.1.2.10), (S)-oxynitrilase (EC 4.1.2.11), and acetone cyanohydrin lyase (EC 4.1.2.37).

The best studied example of the first class is the (R)-oxynitrilase from almonds (*Prunus amygdalus*), or mandelonitrile lyase, which is a monomeric enzyme of 75 kDa, contains a structurally crucial flavine cofactor [502], and is heavily N-glycosylated. The enzyme has been purified [503–505] and crystallized for X-ray studies [506], and the primary structure of a highly homologous enzyme has been deduced from its cDNA [507]. Representatives for the second class are the (S)-oxynitrilases which have been purified for preparative work from millet (*Sorghum bicolor*) [505, 508], and from the leaves of the rubber plant (*Hevea brasiliensis*) [509] and that of *Ximenia americana* [510]. The *Sorghum* oxynitrilase is a glycosylated non-flavoprotein of 95 kDa which has a quaternary $\alpha_2\beta_2$ heteromeric structure consisting of subunits of 25 and 30 kDa [510, 511]. A cDNA clone has been obtained comprising a major part of the genomic sequence encoding the (S)-oxynitrilase subunits [512]. Examples for the third type are oxynitrilases which have been purified and characterized from flax (*Linum usitatissimum*) [513, 514] and tropical cassava leaves (*Manihot esculenta* Crantz), the latter of which has also been cloned [515]. The (R)-(almond) and (S)-oxynitrilases (*Sorghum*) are commercially available.

In general, the method of enzymatic cyanohydrin synthesis promises to be of considerable value in asymmetric synthesis because of the synthetic potential offered by the rich chemistry of enantiomerically pure cyanohydrins, including their stereoselective conversion into other classes of compounds such as α-hydroxy carboxylic acids or respective esters, *vic*-diols, β-aminoalcohols, aziridins, α-azido(amino/fluoro)nitriles, and acyloins [501, 516].

Initial preparative work with oxynitrilases in neutral aqueous solution [517, 518] was hampered by the fact that under these reaction conditions the enzymatic addition has to compete with a spontaneous chemical reaction which limits enantioselectivity. Major improvements in optical purity of cyanohydrins were achieved by conducting the addition under acidic conditions to suppress the uncatalyzed side reaction [519], or by switching to a water immiscible organic solvent as the reaction medium [520], preferably diisopropyl ether. For the latter case, the enzymes are readily immobilized by physical adsorption onto cellulose. A continuous process has been developed for chiral cyanohydrin synthesis using an enzyme membrane reactor [61]. Acetone cyanhydrin can replace the highly toxic hydrocyanic acid as the cyanide source [521]. Inexpensive defatted almond meal has been found to be a convenient substitute for the purified (R)-oxynitrilase without sacrificing enantioselectivity [522–524]. Similarly, lyophilized and powered *Sorghum bicolor* shoots have been successfully tested as an alternative source for the purified (S)-oxynitrilase [525].

The *Sorghum* (S)-oxynitrilase exclusively catalyzes the addition of hydrocyanic acid to aromatic aldehydes with high enantioselectivity, but not to aliphatic aldehydes or ketones [519, 526]. In contrast, the *Hevea* (S)-oxynitrilase was also found to convert aliphatic and α,β-unsaturated substrates with medium to high selectivity [509, 527]. The stereocomplementary almond (R)-oxynitrilase likewise has a very broad substrate tolerance and accepts both aromatic, aliphatic, and α,β-unsaturated aldehydes [520, 521, 523, 528, 529] as well as methyl ketones [530] with high enantiomeric excess (Table 9). It is interesting to note that this enzyme will also tolerate sterically hindered substrates such as pivalaldehyde and suitable derivatives **164** which are effective precursors for (R)-pantolactone **165** [531].

Table 9. Substrate tolerance of hydroxynitrile lyases

R	(R)-oxynitrilase (*Prunus amygdalus*)			(S)-oxynitrilase (*Sorghum bicolor*)			(S)-oxynitrilase (*Hevea brasiliensis*)		
	ee (%)	Yield (%)	Ref.	ee (%)	Yield (%)	Ref.	ee (%)	Yield (%)	Ref.
C_6H_5	99	90	[528]	97	91	[519, 526]	94	79	[509]
$4\text{-}CH_3\text{-}C_6H_4$	98	75	[528]	78	61	[526]	–	–	
$4\text{-}Cl\text{-}C_6H_4$	97	94	[528]	54	87	[526]	–	–	
$3\text{-}C_6H_5O\text{-}C_6H_4$	98	88	[528]	96	93	[526]	20	–	[509]
2-furyl	98	88	[520]	80	80	[501]	–	–	
$E\text{-}CH_3\text{-}CH{=}CH$	99	73	[520, 522, 524]	–	–		86	57	[527]
$n\text{-}C_3H_7$	96	75	[522, 528]	–	–		80	–	[509]
$n\text{-}C_5H_{11}$	94	–	[522, 523]	–	–		84	75	[509]
$(CH_3)_3C$	92	58	[521]	–	–		67	–	[509]

164   **165**

7 Applications to Stereodivergent Asymmetric Synthesis

The availability and the rather similar substrate tolerances exhibited by most members within the families of related lyases fulfills the demand for a complete set of catalysts for a directed stereodivergent synthesis of a set of diastereomeric products from common synthetic building blocks [189]. The verification of this concept ("combinatorial synthesis") has indeed been pursued in a different respect, and will be highlighted by pertinent examples for the synthesis of simple sugars and for the synthesis of 'azasugars', an emerging class of glycosidase inhibitors.

7.1 Monosaccharides

Starting from the enantiomorphous forms of glyceraldehyde, the addition of dihydroxyacetone phosphate, catalyzed by the four available DHAP aldolases FruA, TagA, FucA and RhuA, formally produces each of the 8 possible diastereomeric ketohexoses selectively (Scheme 23). In practice, high yields of product can easily be achieved because both glyceraldehydes are good substrates [195, 202]. Stereoselectivity is complete because of their substrate nature as 2-hydroxyaldehydes, except in the case of the TagA enzyme which is unsuited for conversions of unphosphorylated substrate analogs (Sect. 5.3) [233, 328]. The less common L-ketohexoses produced by RhuA and FucA, L-fructose and L-tagatose, may even be obtained directly from racemic glyceraldehyde by exploiting the capacity of these catalysts for efficient kinetic enantiomer selectivity [347]. Alternatively, highly optically enriched glyceraldehydes may be obtained by Sharpless asymmetric dihydroxylation of the suitably protected acrolein derivative **166** [532].

166

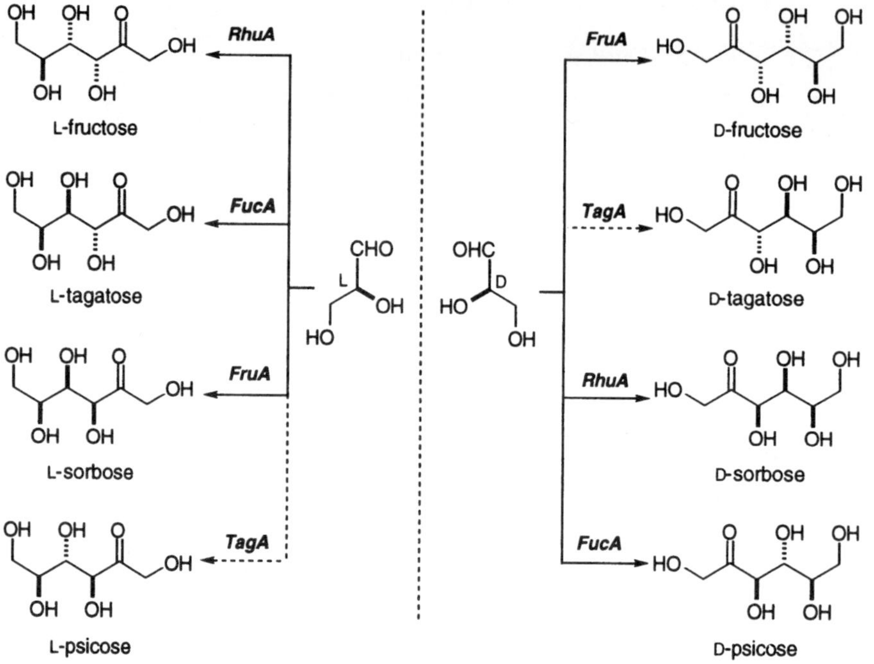

Scheme 23. Stereodivergent synthesis of all eight possible hexoketoses from glyceraldehyde by DHAP aldolases

For a full realization of the concept of stereodivergent carbohydrate synthesis it is necessary to have access to the biologically even more important and structurally more diverse aldose isomers. Interconversion of ketoses and aldoses occurs in Nature by way of enzymatic isomerization for which a number of activities have been reported in the biochemical literature. Both the stereochemically complementary L-rhamnose (RhaI; EC 5.3.1.14) and L-fucose isomerases (FucI; EC 5.3.1.3) from *E. coli* were recently cloned [343, 353, 354, 533] and overexpressed [348]; it has been shown that they possess a relaxed substrate tolerance for sugars and their derivatives having a common (3R) stereochemistry [189, 348]. Thus, by employing the respective RhuA and FucA products (after dephosphorylation), which share the (3R) specificity, in ketose → aldose isomerizations, a broad segment out of the total configurational space of the aldoses becomes accessible in a stereospecific building block manner (Schemes 23 and 24) [190]. Similar results have been realized for the isomerization of D-fructose to D-glucose and a number of derivatives by utilizing a glucose isomerase (GlcI; EC 5.3.1.5) [306], an industrially important enzyme that has a more narrow specificity. In this respect, it is informative that of the four isomeric fluoro ketoses **167–170**, produced by the FruA with the 2- and 3-fluoro analogs **171/172** of glyceraldehyde, the *fructo*-configurated sugar **168**, modified at C-6, was found to be the exclusive substrate of the GlcI [306, 534, 535], while

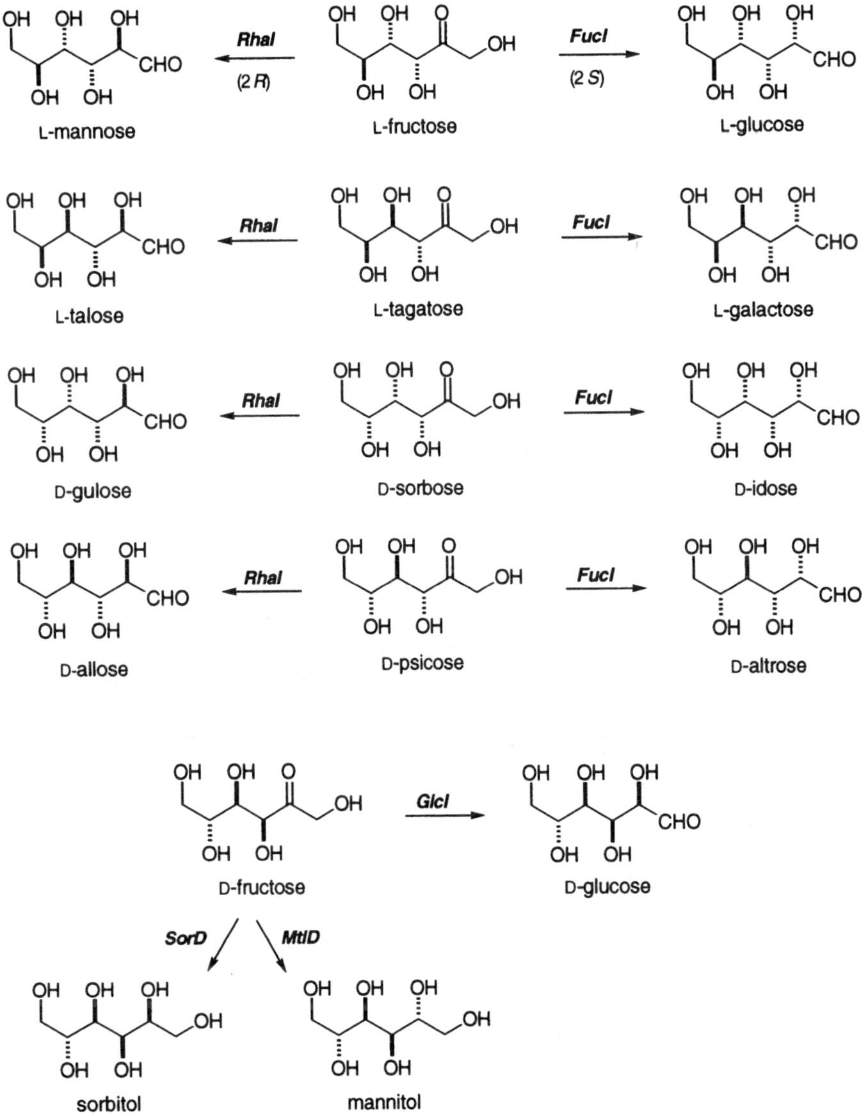

Scheme 24. Stereodivergent isomerizations of hexoketoses by rhamnose and fucose isomerases. Possible enzymatic pathways for the differentiation of fructose

a 5-fluoro substituent (**169, 170**) or the sorbose configuration (**167, 169**) were not allowed [348]. In a search for further structural diversification, a number of FruA products have been investigated as potential substrates for the stereochemically complementary sorbitol dehydrogenase (EC 1.1.1.14) [290] or mannitol dehydrogenase (EC 1.1.1.67) [189, 348]. Indeed, stereospecific reduction to the corresponding alditols could be demonstrated for simple derivatives because of their structural resemblance to D-fructose.

The stereodivergent action of DHAP aldolases is also preparatively useful under conditions of thermodynamic control [202, 230, 294]. Starting from racemic 3-hydroxybutanal **84** the enantiocomplementary nature of the FruA–RhuA enzyme couple will extend to the selection of the dependent stereocenter at C-6 to yield the corresponding mirror imaged products **78** with high selectivity [189]. Conversely, the FucA will contribute a different major diastereomer **173**, albeit only at a low *de* because of more balanced energetic relations [362].

An interesting problem in stereoisomerism is found in the aldol reactions of the achiral aldehydes which are obtained by ozonolysis of the homoallylic alcohols **174**. After stereospecific conversion by the FruA [230], the products can be readily induced to form an intramolecular glycoside **175** by acidic (R=OH) or alkaline treatment (R=Cl), under which conditions the two equatorial ring hydroxyl groups completely direct the stereogenic acetal formation [234]. The corresponding RhuA catalyzed reactions deliver the enantiomeric

176 (R=OH). With less polar chloride substituents, however, the RhuA dia-stereoselectivity is reduced and a considerable fraction of the FucA configurated product (40%) is also formed. Interestingly, alkaline cyclization in the latter occurs with an inverse preference to furnish **177**, which contains the enan-tiomeric bicyclic [3.2.1] structure shared by the FruA product **175** as well as by (S)-(−)-frontalin **178**, the aggregation pheromone of the southern pine beetle *Dendroctonus frontalis*.

In the context of the development of enzymatic methods for the synthesis of higher-carbon sugars we recently developed a new strategy by which dodeca-2,11-diuloses — formally derived from ketohexoses by C-coupling tail to tail — become accessible with deliberately addressable substitution patterns by the twofold aldolase-catalyzed chain elongation ("tandem" aldolization) of simple, readily available dialdehydes [207]. The choice of furanoid (**179**) or pyranoid (**180**) nature of the products can be determined by a suitable placement of hydroxyl substituents in the allylic (**181**) or homoallylic (**182**) positions of a corresponding cycloolefinic precursor.

A number of related acyclic dihydroxy-α,ω-dienes **183** have been successfully evaluated as alternative precursors to such disaccharide mimics for which there is a rather flexible choice in the length of the interlinking chain [234]. Typically, the thermodynamic advantage favors an equatorial attachment of the sugar ring by far, so that the C_2-symmetrical diastereomers can be obtained selectively even when starting from a diol mixture containing both the *rac* and *meso* components.

(major diastereomers)

7.2 Glycosidase Inhibitors

Many alkaloids contain partial structures similar to pyranoses and furanoses in which an imino group replaces the ring oxygen atom. The potent inhibition of exoglycosidases displayed by these compounds has fostered the search for effective syntheses of so-called (1-deoxy) 'azasugars' with a potent and selective inhibition profile for potential therapeutic applications. An important strategy has emerged which consists of consecutive enzymatic aldolizations of suitable azido aldehydes followed by intramolecular reductive amination of the resultant azidoketoses. Since this topic has been amply reviewed in the past [536, 537], only a few examples will be included to provide an outline of the synthetic power of the stereodivergent approach.

Particularly noteworthy are the chemoenzymatic syntheses of diastereomers of the nojirimycin type that have been developed independently by several

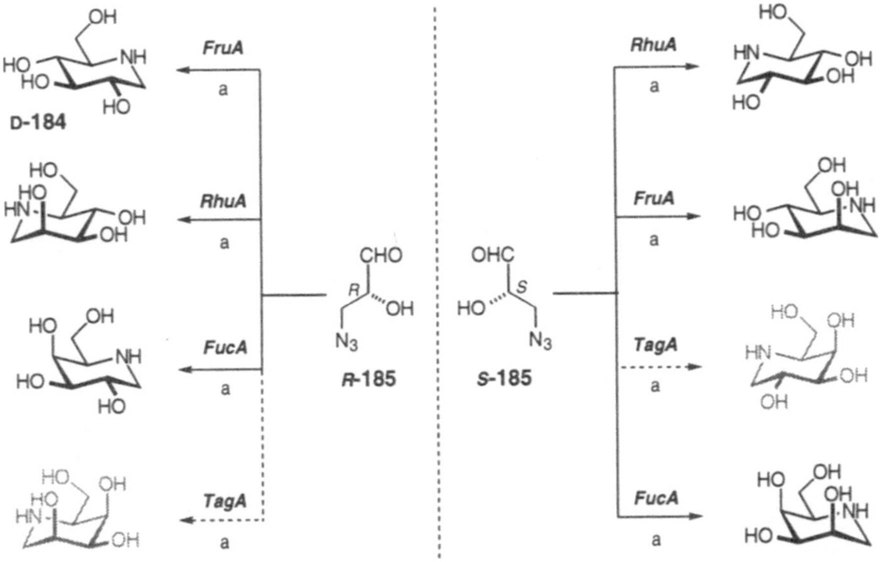

Scheme 25. Stereodivergent synthesis of 1-deoxy azasugars of the nojirimycin type by enzymatic aldolization followed by catalytic reductive amination. a) 1. Aldolase, DHAP; 2. phosphatase; 3. H_2/Pd–C

groups (Scheme 25) [209, 300, 419, 538–541]. For the parent compound 1-deoxy-D-nojirimycin **184**, DHAP was added by FruA catalysis to 3-azido-glyceraldehyde **185**, readily available as a racemate or as pure enantiomers, to provide the respective azidoketose phosphates stereospecifically [419, 538]. Reduction by hydrogenolysis smoothly proceeded via the corresponding amines to the intramolecular Schiff bases which were further reduced stereoselectively to the desired azasugar. In a similar manner the corresponding RhuA and FucA products have been obtained [539–541], with an added benefit that the compounds of the L-series can be made from the racemic aldehyde precursor [200] due to the high kinetic selectivity of these aldolases for the (S)-configurated enantiomer [347]. A less densely functionalized piperidine derivative **186** has been made by acetaldehyde addition to D-**185**, employing the deoxyribose 5-phosphate aldolase [370].

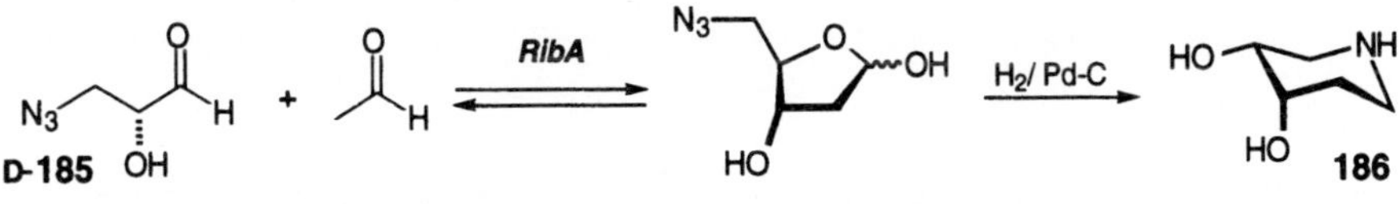

Pyrrolidine-type azasugars can be prepared analogously but starting from the isomeric 2-azidoglyceraldehyde **187** [209, 539, 542]. The five-membered ring compounds are similarly powerful glycosidase inhibitors and are believed to better mimic the flattened-chair transition state of glycosidic cleavage.

Since the usual azasugars lack any substituents at C-1, they show little selectivity for inhibition of α-versus β-glycosidases and only little inhibition towards endoglycosidases which require extended recognition motifs for substrate binding. It has thus been proposed that homologues of azasugars containing an extra hydroxymethylene group at the "anomeric" carbon may enhance potency and/or specificity of inhibition [543]. Towards this end, several C_4 azido-aldehydes have been converted by enzymatic aldolization to the desired homologous azasugars. With the racemic cis-2-azido isomer 188, quite unexpectedly a reversal of the diastereoselectivity for the individual enantiomers has been noted [348]; also a less than desirable selectivity upon catalytic hydrogenation provided both pyrrolidine type azasugars 189 and 190 as mixtures of diastereomers (2:1 and 3:1). For the synthesis of the corresponding piperidine-based isomers, regioisomeric cis-(192) and trans-3-azido aldehydes (191) have been subjected to the same procedure using FruA catalysis, and high stereoselectivity was reported for the reduction steps except for 193 which was obtained as a 1:1 mixture [544, 545].

Addition of pyruvate to Cbz-protected D-mannosamine 194 under NeuA catalysis has furnished an N-acyl derivative of neuraminic acid 5 from which internal reductive amination yielded an azasugar which could be further elaborated to 195, an analog of the bicyclic, indolizidine type glycosidase inhibitor castanospermine [91].

194   **5**   **195**

Two differently *N*-acyl protected derivatives of (*S*)-serinal **197**, readily available from D-glucosamine, have been subjected to aldolizations with DHAP aldolases to furnish the corresponding 5-aminoketoses **196/198** [220, 235]. The latter compounds are interesting in their own right or as further potential precursors for pyrrolidine-type azasugars.

196   **197** HNR   **198**

A very recent application of the tandem aldolization approach towards disaccharide mimics has provided a first example of a *C*-glycosidically linked azadisaccharide **200**. Ozonolysis of the azido substituted cyclohexenediol **199** was followed by a tandem addition of two equivalents of DHAP to the aldehydic termini to yield the corresponding azido substituted dipyranoid 2,11-diulose which, when hydrogenated over a Pd catalyst, highly selectively gave the aza-*C*-disaccharide **200** in 20% overall yield as a single diastereomer [348]. Again, thermodynamic resolution could be advantageously applied to discriminate the racemic starting material **199**.

± **199**   **200**

Thiosugars are another structural variation of carbohydrates that possess interesting biological properties such as the inhibition of glycosidases and related effects. After addition of DHAP to thiosubstituted aldehydes by aldolase catalysis the reaction products immediately cyclize to the stable cyclic thioketose structures. From 2-thioglycolaldehyde **202**, the enantiomeric 5-thio-D-*threo*-pentulofuranoses **201** have been obtained by FruA and RhuA catalysis [208, 420] while a set of thiopyranoses (**204, 205**) has been synthesized from (*R*)-3-thioglyceraldehyde **203** [420, 546]. It is worth noting that with all enzymes an unbiased stereoselectivity is achieved, indicating that the OH and SH substituents are equivalent as regards a correct substrate recognition.

201   **202**   *ent*-**201**

8 Summary and Outlook

A large number of C–C bond forming aldolases have been shown to be of practical value as catalysts for the convergent asymmetric synthesis of interesting polyfunctionalized products. The exemplary reactions that have been carried out with most of the currently available lyases and have been discussed above, will form a useful basis from which the scope and certain limitations of the technique can be explored, and will provide assistance for the evaluation of new potential applications. Clearly, the most notable advantages rely on the high asymmetric induction achievable with biocatalysts and the high level of reaction specificity, which allows for the highly selective transformation of polyfunctionalized substrates without the need for tedious protecting-group manipulations [24]. This includes the opportunity of developing highly integrated reaction schemes, comprising numerous chemical steps of different kinds ("artificial metabolisms") [230, 547]. Whereas new methods for the enzymatic synthesis of many natural or artificial compounds will continue to be developed using the enzymes discussed in this review, it can be expected that further examination of the substrate tolerance of less frequently studied lyases [42], and the exploration of hitherto unknown lyases from a wide range of organisms will be rewarded by an increase in flexibility of this synthetic approach and a broadening of the scope of applications.

Journals concerned with biological disciplines abound with reports on the detection, isolation, characterization and cloning of new and novel enzymes from a diversity of organisms; dozens of new ones are probably added every week to the wealth of our knowledge. Complete genetic information on entire organisms has already been unravelled [548] and more may be expected at an increasing speed thanks to the availability of advanced sequencing and high-speed computing technologies [549]. The goal of altering the substrate specificities for enzymes of different classes is currently being pursued, and the first successful cases have been reported [550–553], including those aimed at a change of cofactor specificity [554, 555]. Thus, a directed modification of substrate specificity and/or stereoselectivity on the basis of detailed structural knowledge or new selection techniques for improved biocatalysts is either already practical or within reach. These and other exciting developments (e.g. in the generation of catalytic antibodies for C–C bond formation [556]) will thus contribute to and hasten the further elaboration of enzymatic techniques, making them a powerful supplement to the existing methodologies of organic synthesis.

Acknowledgements. Our own research in this area has been generously supported by the *Deutsche Forschungsgemeinschaft*, the *Bundesministerium für Bildung, Wissenschaft, Forschung und Technologie*, the *Fonds der Chemischen Industrie, Boehringer Mannheim GmbH* and *Bayer AG*, the *Wissenschaftliche Gesellschaft Freiburg*, and by graduate scholarships administered by the *Ministerium für Wissenschaft und Kunst, Baden Württemberg*. We are grateful to Dr. M. Lawrence (Biomolecular Research Institute, Parkville, Australia) for providing the structural coordinates for the NeuA, and to Dr. M. Dreyer (University of Freiburg, Germany) for providing Fig. 13 and for his assistance with the protein graphics.

Note added in proof:

After finalization of this review, an independent mechanistic model for NeuA catalysis has been proposed (Fitz, W, Schwark, J-R, Wong, C-H (1995) J Org Chem 60:3663) which is based on consideration of products derived from additions to aldotetroses rather than our conformational analysis and rationalization of a three-point binding motif for the natural acceptor substrate (Sect. 3.1, Scheme 2).

9 References

1. Testa B, Trager WF (1990) Chirality 2: 129
2. Wainer IW, Drayer DE (1988) Drug Stereochemistry. Marcel Dekker, New York
3. Ariens EJ, van Rensen JJS, Welling W (1988) Stereoselectivity of Pesticides. Elsevier, Amsterdam
4. Whitesides GM, Wong C-H (1985) Angew Chem Int Ed Engl 24: 617
5. Jones JB (1986) Tetrahedron 42: 3351
6. Davies HG, Green RH, Kelly DR, Roberts SM (1989) Biotransformations in Preparative Organic Chemistry. Academic Press, London
7. Crout DHG, Christen M (1989) Biotransformations in Organic Synthesis In: Scheffold R (ed) Modern Synthetic Methods. Springer, Berlin Heidelberg, New York, vol. 5, p 1
8. Gerhartz W (1991) Enzymes in Industry: Production and Applications. VCH, Weinheim
9. Boland W, Froessl C, Lorenz M (1991) Synthesis 12: 1049
10. Schellenberger V, Jakubke HD (1991) Angew Chem Int Ed Engl 30: 1437
11. Holland HL (1992) Organic Synthesis with Oxidative Enzymes. VCH, Weinheim
12. David S, Augé C, Gautheron C (1991) Adv Carbohydr Chem Biochem 49: 175
13. Blanch HW, Clark DS (1991) Applied Biocatalysis, vol 1. Marcel Dekker, New York
14. Drauz K, Waldmann H (1995) Enzyme Catalysis in Organic Synthesis. A Comprehensive Handbook. VCH, Weinheim

15. Rozzell D, Wagner F (1992) Biocatalytic Production of Amino Acids & Derivatives. Hanser, Munich
16. Wong C-H, Whitesides GM (1994) Enzymes in Synthetic Organic Chemistry. Pergamon, Oxford
17. Faber K (1995) Biotransformations in Organic Chemistry, 2nd edn. Springer, Berlin
18. Servi S (1992) Microbial Reagents in Organic Synthesis, vol 381. Kluwer Academic, Dordrecht
19. Halgas J (1992) Biocatalysts in Organic Synthesis. Elsevier, Amsterdam
20. Poppe L, Novák L (1992) Selective Biocatalysis. A Synthetic Approach. VCH, Weinheim
21. Toone EJ, Simon ES, Bednarski MD, Whitesides GM (1989) Tetrahedron 45: 5365
22. Azerad R (1995) Bull Soc Chim Fr 132: 17
23. Santaniello E, Ferraboschi P, Grisenti P, Manzocchi A (1992) Chem Rev 92: 1071
24. Waldmann H, Sebastian D (1994) Chem Rev 94: 911
25. David S, Auge C, Gautheron C (1992) Adv Carbohydr Chem Biochem 49: 175
26. Varki A (1994) Proc Natl Acad Sci USA 91: 7390
27. Carlos TM, Harlan JM (1994) Blood 84: 2068
28. Hynes RO, Lander AD (1992) Cell 68: 303
29. Wegner CD (1994) Adhesion Molecules. Academic Press, London
30. Chibata I (1978) Immobilized Enzymes. Research and Development. Wiley, New York
31. Trevan MD (1980) Immobilized Enzymes. An Introduction and Applications in Biotechnology. Wiley, New York
32. Mosbach K (1976) Methods Enzymol 44, (1987) 135, (1987) 136, (1988) 137
33. Woodley JM (1992) Immobilized Biocatalysts. In: Smith K (ed) Solids Supports and Catalysts in Organic Synthesis. Ellis Horwood, Chichester, p 254
34. Chenault HK, Whitesides GM (1987) Appl Biochem Biotechnol 14: 147
35. Chenault HK, Simon ES, Whitesides GM (1988) Biotechnol Genet Eng Rev 6: 221
36. Schreiber SL, Verdine GL (1991) Tetrahedron 47: 2543
37. Hedstrom L (1994) Curr Opin Struct Biol 4: 608
38. Janda KD (1994) Proc Natl Acad Sci USA 91: 10779
39. Gordon EM, Barrett RW, Dower WJ, Fodor SPA, Gallop MA (1994) J Med Chem 37: 1385
40. Gallop MA, Barrett RW, Dower WJ, Fodor SPA, Gordon EM (1994) J Med Chem 37: 1233
41. Kluger R (1990) Chem Rev 90: 1151
42. Schomburg D, Salzmann M (1990–1995) Enzyme Handbook. Springer, Berlin
43. Horecker BL, Tsolas O, Lai CY (1972) Aldolases. In: Boyer PD (ed) The Enzymes. 3 edn. Academic Press, New York, vol. VII, p 213
44. Trombetta G, Balboni G, di Iasio A, Grazi E (1977) Biochem Biophys Res Commun 74: 1297
45. Kuo DJ, Rose IA (1985) Biochemistry 24: 3947
46. Grazi E, Cheng T, Horecker BL (1962) Biochem Biophys Res Commun 7: 250
47. Kadonaga JT, Knowles JR (1983) Biochemistry 22: 130
48. Webb EC (1992) Enzyme Nomenclature. Academic Press, San Diego
49. Demerec M, Adelberg EA, Clark AJ, Hartman PE (1966) Genetics 54: 61
50. Smith HO, Nathans D (1973) J Mol Biol 81: 419
51. Kluger R (1990) Chem Rev 90: 1151
52. Littlechild JA, Watson HC (1993) Trends Biochem Sci 18: 36
53. Morris AJ, Tolan DR (1993) J Biol Chem 268: 1095
54. Morris AJ, Tolan DR (1994) Biochemistry 33: 12291
55. Pollak A, Blumenfeld H, Wax M, Baughn RL, Whitesides GM (1980) J Am Chem Soc 102: 6324
56. Bossow B, Berke W, Wandrey C (1992) BioEngineering 8: 12
57. Sobolov SB, Bartoszkomalik A, Oeschger TR, Montelbano MM (1994) Tetrahedron Lett 35: 7751
58. Goße C, Fessner W-D, unpublished
59. Kragl U, Gödde A, Wandrey C, Lubin N, Augé C (1994) J Chem Soc Perkin Trans I 119
60. Kragl U, Gygax D, Ghisalba O, Wandrey C (1991) Angew Chem Int Ed Engl 30: 827
61. Kragl U, Niedermeyer U, Kula M-R, Wandrey C (1990) Ann N Y Acad Sci 613: 167
62. Raetz CRH, Dowhan W (1990) J Biol Chem 265: 1235
63. Unger FM (1981) Adv Carbohydr Chem Biochem 38: 323
64. Varki A (1993) Glycobiology 3: 97
65. Schauer R, Wember M, Wirtz-Peitz F, Ferreira do Amaral C (1971) Hoppe Seyler's Z Physiol Chem 352: 1073
66. Comb DG, Roseman S (1960) J Biol Chem 235: 2529

67. Deijl CM, Vliegenthart JFG (1983) Biochem Biophys Res Commun 111: 668
68. Baumann W, Freidenreich J, Weisshaar G, Brossmer R, Friebolin H (1989) Biol Chem Hoppe Seyler 370: 141
69. Uchida Y, Tsukada Y, Sugimori T (1984) J Biochem 96: 507
70. Ohta Y, Shimosaka M, Murata K, Tsukada Y, Kimura A (1986) Appl Microbiol Biotechnol 24: 386
71. Aisaka K, Tamura S, Arai Y, Uwajima T (1987) Biotechnol Lett 9: 633
72. Ohta Y, Tsukada Y, Sugimori T, Murata K, Kimura A (1989) Agric Biol Chem 53: 477
73. Aisaka K, Igarashi A, Yamaguchi K, Uwajima T (1991) Biochem J 276: 541
74. Izard T, Lawrence MC, Malby RL, Lilley GG, Colman PM (1994) Structure 2: 361
75. Schauer R, Stoll S, Zbiral E, Schreiner E, Brandstetter HH, Vasella A, Baumberger F (1987) Glycoconjugate J 4: 361
76. Kraulis PJ (1991) J Appl Crystallogr 24: 946
77. Kim MJ, Hennen WJ, Sweers HM, Wong C-H (1988) J Am Chem Soc 110: 6481
78. Harlan JM, Liu DY (1992) Adhesion – Its Role in Inflammatory Disease. W.H. Freeman and Co, New York
79. Kelm S, Schauer R, Manuguerra JC, Gross HJ, Crocker PR (1994) Glycoconjugate J 11: 576
80. Troy FA (1992) Glycobiology 2: 5
81. Wiley DC, Skehel JJ (1987) Annu Rev Biochem 56: 365
82. Augé C, David S, Gautheron C, Malleron A, Cavayé B (1988) New J Chem 12: 733
83. Simon ES, Bednarski MD, Whitesides GM (1988) J Am Chem Soc 110: 7159
84. Lin CH, Sugai T, Halcomb RL, Ichikawa Y, Wong C-H (1992) J Am Chem Soc 114: 10138
85. Ichikawa Y, Liu JLC, Shen GJ, Wong C-H (1991) J Am Chem Soc 113: 6300
86. Blayer S, Woodley JM, Dawson MJ, Lilly MD (1995) Process Development for Scale-Up of Biotransformations: Chemoenzymatic Synthesis of N-Acetylneuraminic Acid. Biotrans 95, Warwick, UK
87. Liu JLC, Shen GJ, Ichikawa Y, Rutan JF, Zapata G, Vann WF, Wong C-H (1992) J Am Chem Soc 114: 3901
88. Augé C, Gautheron C, David S, Malleron A, Cavayé B, Bouxom B (1990) Tetrahedron 46: 201
89. Koppert K, Brossmer R (1992) Tetrahedron Lett 33: 8031
90. Sparks MA, Williams KW, Lukacs C, Schrell A, Priebe G, Spaltenstein A, Whitesides GM (1993) Tetrahedron 49: 1
91. Zhou PZ, Salleh HM, Honek JF (1993) J Org Chem 58: 264
92. Fitz W, Wong C-H (1994) J Org Chem 59: 8279
93. Gross HJ, Brossmer R (1988) Eur J Biochem 177: 583
94. Brossmer R, Gross HJ (1994) Methods Enzymol 247: 177
95. Brossmer R, Gross HJ (1994) Methods Enzymol 247: 153
96. Jacob GS, Kirmaier C, Abbas SZ, Howard SC, Steininger CN, Welply JK, Scudder P (1995) Biochemistry 34: 1210
97. Varki A (1992) Glycobiology 2: 25
98. Halcomb RL, Fitz W, Wong C-H (1994) Tetrahedron Asymmetry 5: 2437
99. Kong DCM, von Itzstein M (1995) Tetrahedron Lett 36: 957
100. Isecke R, Brossmer R (1994) Tetrahedron 50: 7445
101. Schrell A, Whitesides GM (1990) Liebigs Ann Chem 1111
102. Kragl U, Gödde A, Wandrey C, Kinzy W, Cappon JJ, Lugtenburg J (1993) Tetrahedron Asymmetry 4: 1193
103. Augé C, Bouxom B, Cavayé B, Gautheron C (1989) Tetrahedron Lett 30: 2217
104. David S, Malleron A, Cavayé B (1992) New J Chem 16: 751
105. Gautheron LNC, Ichikawa Y, Wong C-H (1991) J Am Chem Soc 113: 7816
106. Ladisch S, Hasegawa A, Li RX, Kiso M (1995) Biochemistry 34: 1197
107. Knappmann BR, Kula MR (1990) Appl Microbiol Biotechnol 33: 324
108. Ghalambor MA, Heath EC (1966) J Biol Chem 241: 3222
109. Sugai T, Shen GJ, Ichikawa Y, Wong C-H (1993) J Am Chem Soc 115: 413
110. Hammond SM, Claesson A, Jansson AM, Larsson L-G, Pring BG, Town CM, Ekström B (1987) Nature 327: 730
111. Claesson A (1987) J Org Chem 52: 4414
112. Wood WA (1972) 2-Keto-3-deoxy-6-phosphogluconic and Related Aldolases. In: Boyer PD (ed) The Enzymes, 3rd ed. Academic, New York, vol. 7, p 281
113. Scopes RK (1984) Anal Biochem 136: 525
114. Shelton MC, Toone EJ (1995) Tetrahedron Asymmetry 6: 207

115. Knappmann BR, Elnawawy MA, Schlegel HG, Kula MR (Erratum (1995) Carbohydr Res 272: C1)
116. Egan SE, Fliege R, Tong S, Shibata A, Wolf RE, Conway T (1992) J Bacteriol 174: 4638
117. Carter AT, Pearson BM, Dickinson JR, Lancashire WE (1993) Gene 130: 155
118. Conway T, Yi KC, Egan SE, Wolf REJ, Rowley DL (1991) J Bacteriol 173: 5247
119. Conway T, Fliege R, Jones-Kilpatrick D, Liu J, Barnell WO, Egan SE (1991) Mol Microbiol 5: 2901
120. Hugouvieux-Cotte-Pattat N, Robert-Baudouy J (1994) Mol Microbiol 11: 67
121. Suzuki N, Wood WA (1980) J Biol Chem 255: 3427
122. Hammerstedt RH, Möhler H, Decker KA, Ersfeld D, Wood WA (1975) Methods Enzymol 42: 258
123. Mavridis IM, Hatada MH, Tulinsky A, Lebioda L (1982) J Mol Biol 162: 419
124. Dekker EE, Kitson RP (1992) J Biol Chem 267: 10507
125. Dekker EE, Kobes RD, Grady SR (1975) Methods Enzymol 42: 280
126. Dekker EE, Nishihara H, Grady SR (1975) Methods Enzymol 42: 285
127. Kuratomi K, Fukunaga K (1963) Biochim Biophys Acta 78: 617
128. Maitra U, Dekker EE (1964) J Biol Chem 239: 1485
129. Rosso RG, Adams E (1967) J Biol Chem 242: 5524
130. Vlahos CJ, Dekker EE (1988) J Biol Chem 263: 11683
131. Patil RV, Dekker EE (1992) J Bacteriol 174: 102
132. Taha TSM, Deits TL (1994) Biochem Biophys Res Commun 200: 459
133. Meloche HP, Mehler L (1973) J Biol Chem 248: 6333
134. Meloche HP, Monti CT (1975) J Biol Chem 250: 6875
135. Scholtz JM, Schuster SM (1984) Bioorg Chem 12: 229
136. Floyd NC, Liebster MH, Turner NJ (1992) J Chem Soc Perkin Trans I 1085
137. Allen ST, Heintzelman GR, Toone EJ (1992) J Org Chem 57: 426
138. Nishihara H, Dekker EE (1972) J Biol Chem 247: 5079
139. Meloche HP, O'Connell EL (1982) Methods Enzymol 90: 263
140. Dahms AS, Anderson RL (1972) J Biol Chem 247: 2238
141. Anderson RL, Dahms AS (1975) Methods Enzymol 42: 269
142. Dahms AS (1974) Biochem Biophys Res Commun 60: 1433
143. Dahms AS, Donald A (1982) Methods Enzymol 90: 269
144. Augé C, Delest V (1993) Tetrahedron Asymmetry 4: 1165
145. Elshafei AM, Abdel-Fatah OM (1989) Enzyme Microb Technol 11: 367
146. Elshafei AM, Mohawed SM, Ammar MS, Abdel-Fatah OM (1995) Int J Gen Mol Microbiol 67: 211
147. Auge C, Delest V (1995) Tetrahedron Asymmetry 6: 863
148. Fish DC, Blumenthal HJ (1966) Methods Enzymol 9: 529
149. Hirschbein BL, Mazenod FP, Whitesides GM (1982) J Org Chem 47: 3765
150. Simon ES, Grabowski S, Whitesides GM (1989) J Am Chem Soc 111: 8920
151. Blacklow RS, Warren L (1962) J Biol Chem 237: 3520
152. Annunziato PW, Wright LF, Vann WF, Silver RP (1995) J Bacteriol 177: 312
153. Ganguli S, Zapata G, Wallis T, Reid C, Boulnois G, Vann WF, Roberts IS (1994) J Bacteriol 176: 4583
154. Edwards U, Müller A, Hammerschmidt S, Gerardy-Schahn R, Frosch M (1994) Mol Microbiol 14: 141
155. Brossmer R, Rose U, Kasper D, Smith TL, Grasmuk H, Unger FM (1980) Biochem Biophys Res Commun 96: 1282
156. Raetz CRH (1993) J Bacteriol 175: 5745
157. Doong RL, Ahmad S, Jensen RA (1991) Plant Cell Environ 14: 113
158. Levin DH, Racker E (1959) J Biol Chem 234: 2532
159. Woisetschläger M, Högenauer G (1987) MGG Mol Gen Genet 207: 369
160. Woisetschläger M, Högenauer G (1986) J Bacteriol 168: 437
161. Dotson GD, Dua RK, Clemens JC, Wooten EW, Woodard RW (1995) J Biol Chem 270: 13698
162. Ray PH (1980) J Bacteriol 141: 635
163. Ray PH (1982) Method Enzymol 83: 525
164. Baasov T, Sheffer-Dee-Noor S, Kohen A, Jakob A, Belakhov V (1993) Eur J Biochem 217: 991
165. Yanase H, Okuda M, Kita K, Shibata K, Sakai Y, Kato Y (1995) Appl Microbiol Biotechnol 43: 228

166. Bednarski MD, Crans DC, DiCosimo R, Simon ES, Stein PD, Whitesides GM, Schneider MJ (1988) Tetrahedron Lett 29: 427
167. Wong C-H, Drueckhammer DG, Durrwachter JR, Lacher B, Chauvet CJ, Wang YF, Sweers HM, Smith GL, Yang LJS, Hennen WJ (1989) Enzyme-catalyzed synthesis of carbohydrates. In: Horton D, Hawkins LD, McGarvey GJ (eds) Trends in Synthetic Carbohydrate Chemistry. American Chemical Society, Washington, vol. 386, p 317
168. Doong RL, Jensen RA (1992) New Phytol 121: 165
169. Srinivasan PR, Sprinson DB (1959) J Biol Chem 234: 716
170. Hurwitz J, Weissbach A (1959) J Biol Chem 234: 710
171. Zurawski G, Brown KD, Killingly D, Yanofsky C (1978) Proc Natl Acad Sci USA 75: 4271
172. Shultz J, Hermodson MA, Garner CC, Herrmann KM (1984) J Biol Chem 259: 9655
173. Ray JM, Yanofsky C, Baeuerle R (1988) J Bacteriol 170: 5500
174. Weaver LM, Herrmann KM (1990) J Bacteriol 172: 6581
175. Sprinson DB, Srinivasan PR, Katagiri M (1962) Methods Enzymol 5: 394
176. Schoner R, Herrmann KM (1976) J Biol Chem 251: 5440
177. Paravicini G, Schmidheini T, Braus G (1989) Eur J Biochem 186: 361
178. Ray JM, Bauerle R (1991) J Bacteriol 173: 1894
179. Stuart F, Hunter IS (1993) Biochim Biophys Acta 1161: 209
180. Muday GK, Herrmann KM (1990) J Bacteriol 172: 2259
181. Dyer WE, Weaver LM, Zhao J, Kuhn DN, Weller SC, Herrmann KM (1990) J Biol Chem 265: 1608
182. Weaver LM, Pinto JEBP, Herrmann KM (1993) Bioorg Med Chem Lett 3: 1421
183. Doong RL, Gander JE, Ganson RJ, Jensen RA (1992) Physiol Plant 84: 351
184. Reimer LM, Conley DL, Pompliano DL, Frost JW (1986) J Am Chem Soc 108: 8010
185. Montchamp J-L, Piehler LT, Frost JW (1992) J Am Chem Soc 114: 4453
186. Draths KM, Frost JW (1990) J Am Chem Soc 112: 1657
187. Draths KM, Ward TL, Frost JW (1992) J Am Chem Soc 114: 9725
188. Draths KM, Frost JW (1995) J Am Chem Soc 117: 2395
189. Fessner W-D (1992) A Building Block Strategy for Asymmetric Synthesis: The DHAP Aldolases. In: Servi S (ed) Microbial Reagents in Organic Synthesis. Kluwer Academic, Dordrecht, vol. 381, p 43
190. Fessner W-D (1993) GIT Fachz Lab 37: 951
191. Lebherz HG (1972) Biochemistry 11: 2243
192. Lebherz HG, Rutter WJ (1969) Biochemistry 8: 109
193. Harris CE, Kobes RD, Teller DC, Rutter WJ (1969) Biochemistry 8: 2442
194. Schwartz NB, Abram D, Feingold DS (1974) Biochemistry 13: 1726
195. Fessner W-D, Sinerius G, Schneider A, Dreyer M, Schulz GE, Badia J, Aguilar J (1991) Angew Chem Int Ed Engl 30: 555
196. Taylor JF, Green AA, Cori GT (1948) J Biol Chem 173: 591
197. Vanderheiden BS, Meinhart JO, Dodson RG, Krebs EG (1962) J Biol Chem 237: 2095
198. Richards OC, Rutter WJ (1961) J Biol Chem 236: 3177
199. Periana RA, Motiu-DeGrood R, Chiang Y, Hupe DJ (1980) J Am Chem Soc 102: 3923
200. Fessner W-D, Sinerius G (1994) Angew Chem Int Ed Engl 33: 209
201. Stribling D (1974) Biochem J 141: 725
202. Bednarski MD, Simon ES, Bischofberger N, Fessner W-D, Kim MJ, Lees W, Saito T, Waldmann H, Whitesides GM (1989) J Am Chem Soc 111: 627
203. Whitesides GM, personal communication
204. Paterson MC, Norton IL, Hartman FC (1972) Biochemistry 11: 2070
205. Magnien A, Le Clef B, Biellmann JF (1984) Biochemistry 23: 6858
206. Weis K, Fessner W-D, unpublished
207. Eyrisch O, Fessner W-D (1995) Angew Chem Int Ed Engl 34: 1639
208. Fessner W-D, Sinerius G (1994) Bioorg Med Chem 2: 639
209. von der Osten CH, Sinskey AJ, Barbas CF, Pederson RL, Wang YF, Wong C-H (1989) J Am Chem Soc 111: 3924
210. Götz F, Fischer S, Schleifer K-H (1980) Eur J Biochem 108: 295
211. Dreyer MK, Schulz GE (1993) J Mol Biol 231: 549
212. Sinerius G, Schneider A, Goße C, Fessner W-D, unpublished
213. Rutter WJ, Richards OC, Woodfin BM (1961) J Biol Chem 236: 3193
214. Spolter PD, Adelman RC, Weinhouse S (1965) J Biol Chem 240: 1327

215. Berthiaume L, Tolan DR, Sygusch J (1993) J Biol Chem 268: 10826
216. Berthiaume L, Loisel TP, Sygusch J (1991) J Biol Chem 266: 17099
217. Richard JP (1984) J Am Chem Soc 106: 4926
218. Richard JP (1993) Biochem Soc Trans 21: 549
219. Phillips SA, Thornalley PJ (1993) Eur J Biochem 212: 101
220. Sinerius G (1994) Ph.D. thesis, University of Freiburg
221. Waagen V, Barua TK, Anthonsen HW, Hansen LK, Fossli DJ, Hough E, Anthonsen T (1994) Tetrahedron 50: 10055
222. Effenberger F, Straub A (1987) Tetrahedron Lett 28: 1641
223. Pederson RL, Esker J, Wong C-H (1991) Tetrahedron 47: 2643
224. Colbran RL, Jones JKN, Matheson NK, Rozema I (1967) Carbohydr Res 4: 355
225. Jung SH, Jeong JH, Miller P, Wong C-H (1994) J Org Chem 59: 7182
226. Crans DC, Whitesides GM (1985) J Am Chem Soc 107: 7019
227. Crans DC, Kazlauskas RJ, Hirschbein BL, Wong C-H, Abril O, Whitesides GM (1987) Methods Enzymol 136: 263
228. Meyerhof O, Lohmann K, Schuster P (1936) Biochem Zeitschr 286: 301
229. Willnow P (1984) Fructose-1,6-bisphosphate Aldolase. In: Bergmeyer H-J (ed) Methods of Enzymatic Analysis. 3rd edn. Verlag Chemie, Weinheim, vol. 4, p 346
230. Fessner W-D, Walter C (1992) Angew Chem Int Ed Engl 31: 614
231. Colombo G, Tate PW, Girotti AW, Kemp RG (1975) J Biol Chem 250: 9404
232. Babul J (1978) J Biol Chem 253: 4350
233. Eyrisch O, Sinerius G, Fessner W-D (1993) Carbohydr Res 238: 287
234. Walter C (1996) Ph.D. thesis, RWTH Aachen
235. Liu KKC, Pederson RL, Wong C-H (1991) J Chem Soc, Perkin Trans 1: 2669
236. Kajimoto T, Liu KKC, Pederson RL, Zhong Z, Ichikawa Y, Porco JAJ, Wong C-H (1991) J Am Chem Soc 113: 6187
237. Kajimoto T, Chen L, Liu KKC, Wong C-H (1991) J Am Chem Soc 113: 6678
238. Esders TW, Michrina CA (1979) J Biol Chem 254: 2710
239. Arth H-L, Sinerius G, Fessner W-D (1995) Liebigs Ann Chem 2037
240. Drueckhammer DG, Durrwachter JR, Pederson RL, Crans DC, Daniels L, Wong C-H (1989) J Org Chem 54: 70
241. Gresser MJ, Tracey AS (1990) Vanadates as Phosphate Analogs in Biochemistry. In: Chasteen ND (ed) Vanadium in Biological Systems. Physiology and Biochemistry. Kluwer Academic, Dordrecht, p 63
242. Nour-Eldeen AF, Craig MM, Gresser MJ (1985) J Biol Chem 260: 6836
243. Crans DC, Schelble SM, Theisen LA (1991) J Org Chem 56: 1266
244. London J (1974) J Biol Chem 249: 7977
245. Sugimoto S, Nosoh Y (1971) Biochim Biophys Acta 235: 210
246. Freeze H, Brock TD (1970) J Bacteriol 101: 541
247. Ingram JM, Hochster RM (1967) Can J Biochem 45: 929
248. Stribling D, Perham RN (1973) Biochem J 131: 833
249. Barnes EM, Akagi JM, Himes RH (1971) Biochim Biophys Acta 227: 199
250. Krishnan G, Altekar W (1991) Eur J Biochem 195: 343
251. Baldwin SA, Perham RN (1978) Biochem J 169: 643
252. Brockamp HP, Kula MR (1990) Appl Microbiol Biotechnol 34: 287
253. Lebherz HG, Rutter WJ (1973) J Biol Chem 248: 1650
254. Bai NJ, Pai MR, Murthy PS, Venkitasubramanian TA (1975) Arch Biochem Biophys 168: 235
255. Schnarrenberger C, Pelzer-Reith B, Yatsuki H, Freund S, Jacobshagen S, Hori K (1994) Arch Biochem Biophys 313: 173
256. Russell GK, Gibbs M (1967) Biochim Biophys Acta 132: 145
257. Willard J, Gibbs M (1968) Biochim Biophys Acta 151: 438
258. Fluri R, Ramasarma T, Horecker BL (1967) Eur J Biochem 1: 117
259. Krüger I, Schnarrenberger C (1983) Eur J Biochem 136: 101
260. Lebherz HG, Leadbetter MM, Bradshaw RA (1984) J Biol Chem 259: 1011
261. Heil JA, Lebherz HG (1978) J Biol Chem 253: 6599
262. Valentin ML, Bolte J (1993) Tetrahedron Lett 34: 8103
263. Fernández-Sousa J, Gavilanes FG, Gavilanes JG, Paredes JA (1978) Arch Biochem Biophys 188: 456
264. Brenner-Holzach O (1979) Arch Biochem Biophys 194: 321
265. Kochman M, Kwiatkowska D (1972) Arch Biochem Biophys 152: 856

266. Certa U, Ghersa P, Döbeli H, Matile H, Kocher HP, Shrivastava IK, Shaw AR, Perrin LH (1988) Science 240: 1036
267. Ting S-M, Sia CL, Lai CY, Horecker BL (1971) Arch Biochem Biophys 144: 485
268. Komatsu SK, Feeney R (1970) Biochim Biophys Acta 206: 305
269. Yeltman DR, Harris BG (1977) Biochim Biophys Acta 484: 188
270. Rajkumar TV, Woodfin BM, Rutter WJ (1966) Methods Enzym 9: 491
271. Marquardt RR (1971) Can J Biochem 49: 658
272. Peanasky RJ, Lardy HA (1990) J Biol Chem 265: 1608
273. Gracy RW, Lacko AG, Horecker BL (1969) J Biol Chem 244: 3913
274. Penhoet EE, Kochman M, Rutter WJ (1969) Biochemisty 8: 4391
275. Izzo P, Costanzo P, Lupo A, Rippa E, Paolella G, Salvatore F (1988) Eur J Biochem 174: 569
276. von der Osten CH, Barbas CF, Wong C-H, Sinskey AJ (1989) Mol Microbiol 3: 1625
277. Schwelberger HG, Kohlwein SD, Paltauf F (1989) Eur J Biochem 180: 301
278. Mutoh N, Hayashi Y (1994) Biochim Biophys Acta 1183: 550
279. Berthiaume L, Beaudry D, Lazure C, Tolan DR, Sygusch J (1989) Arch Biochem Biophys 272: 281
280. Alefounder PR, Baldwin SA, Perham RN, Short NJ (1989) Biochem J 257: 529
281. Sygusch J, Beaudry D, Allaire M (1987) Proc Natl Acad Sci USA 84: 7846
282. Gamblin SJ, Davies GJ, Grimes JM, Jackson RM, Littlechild JA, Watson HC (1991) J Mol Biol 219: 573
283. Gamblin SJ, Cooper B, Millar JR, Davies GJ, Littlechild JA, Watson HC (1990) FEBS Lett 262: 282
284. Hester G, Brenner-Holzach O, Rossi FA, Struck-Donatz M, Winterhalter KH, Smit JDG, Piontek K (1991) FEBS Lett 292: 237
285. Brockamp HP, Kula MR (1990) Tetrahedron Lett 31: 7123
286. Walter C, Petersen M, Fessner W-D, unpublished
287. Wong C-H, Whitesides GM (1983) J Org Chem 48: 3199
288. Schmid W, Whitesides GM (1990) J Am Chem Soc 112: 9670
289. Schmid W, Heidlas J, Mathias JP, Whitesides GM (1992) Liebigs Ann Chem 95
290. Borysenko CW, Spaltenstein A, Straub JA, Whitesides GM (1980) J Am Chem Soc 111: 9275
291. Wong C-H, Mazenod FP, Whitesides GM (1983) J Org Chem 48: 3493
292. Durrwachter JR, Sweers HM, Nozaki K, Wong C-H (1986) Tetrahedron Lett 27: 1261
293. Peters J, Brockamp HP, Minuth T, Grothus M, Steigel A, Kula MR, Elling L (1993) Tetrahedron Asymmetry 4: 1173
294. Durrwachter JR, Wong C-H (1988) J Org Chem 53: 4175
295. Schneider A (1990) Diploma thesis, University of Freiburg
296. Jaeschke G, Goße C, Petersen M, Fessner W-D, unpublished
297. Bednarski MD, Waldmann HJ, Whitesides GM (1986) Tetrahedron Lett 27: 5807
298. Fessner W-D, Sinerius G, Schneider A, Weis K (1991) Limits of Aldolase Stereoselectivity. 4th Chemical Congress of North America and 202nd ACS National Meeting, New York, BIOL 9
299. Lees WJ, Whitesides GM (1993) J Org Chem 58: 1887
300. Straub A, Effenberger F, Fischer P (1990) J Org Chem 55: 3926
301. Fessner W-D, unpublished
302. Angyal SJ (1969) Angew Chem Int Ed Engl 8: 157
303. Jones JKN, Sephton HH (1960) Can J Chem 38: 753
304. Kapuscinski M, Franke FP, Flanigan I, MacLeod JK, Williams JF (1985) Carbohydr Res 140: 69
305. Paoletti F, Williams JF, Horecker BL (1979) Arch Biochem Biophys 198: 614
306. Durrwachter JR, Drueckhammer DG, Nozaki K, Sweers HM, Wong C-H (1986) J Am Chem Soc 108: 7812
307. Turner NJ, Whitesides GM (1989) J Am Chem Soc 111: 624
308. Liu KKC, Wong C-H (1992) J Org Chem 57: 4789
309. Maliakel BP, Schmid W (1993) J Carbohydr Chem 12: 415
310. Nicotra F, Panza L, Russo G, Verani A (1993) Tetrahedron Asymmetry 4: 1203
311. Matsumoto K, Shimagaki M, Nakata T, Oishi T (1993) Tetrahedron Lett 34: 4935
312. Shimagaki M, Muneshima H, Kubota M, Oishi T (1993) Chem Pharm Bull 41: 282
313. Schultz M, Waldmann H, Kunz H, Vogt W (1990) Liebigs Ann Chem 1019
314. Myles DC, Andrulis PJI, Whitesides GM (1991) Tetrahedron Lett 32: 4835
315. Bissett DL, Anderson RL (1973) Biochem Biophys Res Commun 52: 641
316. Bissett DL, Anderson RL (1974) J Bacteriol 119: 698

317. Bissett DL, Anderson RL (1974) J Bacteriol 117: 318
318. Hamilton IR, Lebtag H (1979) J Bacteriol 140: 1102
319. Crow VL, Davey GP, Pearce LE, Thomas TD (1983) J Bacteriol 153: 76
320. Kanatani K, Tahara T, Yoshida K, Miura H, Sakamoto M, Oshimura M (1992) Biosci Biotechnol Biochem 56: 826
321. Szumilo T (1981) FEBS Lett 124: 270
322. Markwell J, Shimamoto GT, Bissett DL, Anderson RL (1976) Biochem Biophys Res Commun 71: 221
323. Lengeler J (1977) MGG Mol Gen Genet 152: 83
324. Bissett DL, Anderson RL (1980) J Biol Chem 255: 8750
325. Anderson RL, Bissett DL (1982) Methods Enzymol 90: 228
326. Crow VL, Thomas TD (1982) J Bacteriol 151: 600
327. Anderson RL, Markwell JP (1982) Methods Enzymol 90: 232
328. Fessner W-D, Eyrisch O (1992) Angew Chem Int Ed Engl 31: 56
329. Limsowtin GKY, Crow VL, Pearce LE (1986) FEMS Microbiol Lett 33: 79
330. Yu PL, Limsowtin GKY, Crow VL, Pearce LE (1988) Appl Microbiol Biotechnol 28: 471
331. Van Rooijen RJ, Van Schalkwijk S, De Vos WM (1991) J Biol Chem 266: 7176
332. Rosey EL, Oskouian B, Stewart GC (1991) J Bacteriol 173: 5992
333. Jagusztyn-Krynicka EK, Hansen JB, Crow VL, Thomas TD, Honeyman AL, Curtiss R (1992) J Bacteriol 174: 6152
334. Rosey EL, Stewart GC (1992) J Bacteriol 174: 6159
335. Alpert CA, unpublished
336. Takagi Y (1966) Methods Enzymol 9: 542
337. Chiu TH, Evans KL, Feingold DS (1975) Methods Enzymol 42: 264
338. Chiu T-H, Feingold DS (1969) Biochemistry 8: 98
339. Sawada H, Takagi Y (1964) Biochim Biophys Acta 92: 26
340. Schwartz NB, Feingold DS (1972) Bioinorg Chem 1: 233
341. Schwartz NB, Feingold DS (1972) Bioinorg Chem 2: 75
342. Badía J, Gimenez R, Baldomà L, Barnes E, Fessner W-D, Aguilar J (1991) J Bacteriol 173: 5144
343. Moralejo P, Egan SM, Hidalgo E, Aguilar J (1993) J Bacteriol 175: 5585
344. Nishitani J, Wilcox G (1991) Gene 105: 37
345. Axmann S, Schulz GE, unpublished
346. Fessner W-D, Schneider A, Eyrisch O, Sinerius G, Badia J (1993) Tetrahedron Asymmetry 4: 1183
347. Fessner W-D, Badia J, Eyrisch O, Schneider A, Sinerius G (1992) Tetrahedron Lett 33: 5231
348. Eyrisch O (1994) Ph.D. thesis, University of Freiburg
349. Mazur AW (1991) Galactose Oxidase: Selected Properties and Synthetic Applications. In: Bednarski MD, Simon ES (eds) Enzymes in Carbohydrate Synthesis. American Chemical Society, Washington, vol. 466, p 99
350. Eyrisch O, Keller M, Fessner W-D (1994) Tetrahedron Lett 35: 9013
351. Ghalambor MA, Heath EC (1962) J Biol Chem 237: 2427
352. Ghalambor MA, Heath EC (1966) Methods Enzymol 9: 538
353. Chen YM, Zhu Y, Lin ECC (1987) MGG Mol Gen Genet 210: 331
354. Lu Z, Lin ECC (1989) Nucleic Acids Res 17: 4883
355. Ozaki A, Toone EJ, von der Osten CH, Sinskey AJ, Whitesides GM (1990) J Am Chem Soc 112: 4970
356. Dreyer M, Schulz GE, unpublished
357. Fessner W-D, Schneider A, Held H, Sinerius G, Walter C, Hixon M, Schloss JV (1996) Angew Chem Int Ed Engl 35: 0000
358. Mildvan AS, Kobes RD, Rutter WJ (1971) Biochemistry 10: 1191
359. Smith GM, Mildvan AS, Harper ET (1980) Biochemistry 19: 1248
360. Smith GM, Mildvan AS (1981) Biochemistry 20: 4340
361. Belasco JG, Knowles JR (1983) Biochemistry 22: 122
362. Schneider A (1994) Ph.D. thesis, University of Freiburg
363. Racker E (1952) J Biol Chem 196: 347
364. Hoffee P, Rosen OM, Horecker BL (1966) Methods Enzymol 9: 545
365. Hoffee PA (1968) Arch Biochem Biophys 126: 795
366. Hoffee P, Rosen OM, Horecker BL (1965) J Biol Chem 240: 1512
367. Stura EA, Ghosh S, Garcia-Junceda E, Chen LR, Wong C-H, Wilson IA (1995) Proteins 22: 67

368. Valentin-Hansen P, Boetius F, Hammer-Jespersen K, Svendsen I (1982) Eur J Biochem 125: 561
369. Gijsen HJM, Wong C-H (1995) J Am Chem Soc 117: 2947
370. Chen L, Dumas DP, Wong C-H (1992) J Am Chem Soc 114: 741
371. Barbas CF, Wang YF, Wong C-H (1990) J Am Chem Soc 112: 2013
372. Gijsen HJM, Wong C-H (1994) J Am Chem Soc 116: 8422
373. Wong C-H, Garcia-Junceda E, Chen LR, Blanco O, Gijsen HJM, Steensma DH (1995) J Am Chem Soc 117: 3333
374. Beisswenger R, Kula MR (1991) Appl Microbiol Biotechnol 34: 604
375. Levering PR, Croes LM, Dijkhuizen L (1986) Arch Microbiol 144: 279
376. Yanase H, Matsuzaki K, Sato Y, Kita K, Sato Y, Kato N (1992) Appl Microbiol Biotechnol 37: 301
377. Yanase H, Koike Y, Matsuzaki K, Kita K, Sato Y, Kato N (1992) Biosci Biotechnol Biochem 56: 541
378. Kemp MB (1974) Biochem J 139: 129
379. Brockamp HP, Steigel A, Kula MR (1993) Liebigs Ann Chem 621
380. Beisswenger R, Snatzke G, Thieme J, Kula MR (1991) Tetrahedron Lett 32: 3159
381. Lill U, Pirzer P, Kukla D, Huber R, Eggerer H (1980) Hoppe Seyler's Z Physiol Chem 361: 875
382. Banki K, Halladay D, Perl A (1994) J Biol Chem 269: 2847
383. Schaaf I, Hohmann S, Zimmermann FK (1990) Eur J Biochem 188: 597
384. Williams JF, Blackmore PF (1983) Int J Biochem 15: 797
385. Arora KK, Collins JG, MacLeod JK, Williams JF (1988) Biol Chem Hoppe Seyler 369: 549
386. Horecker BL, Smyrniotis PZ (1953) J Am Chem Soc 75: 2021
387. Moradian A, Benner SA (1992) J Am Chem Soc 114: 6980
388. Grazi E, Mangiarotti M, Pontremoli S (1962) Biochemistry 1: 628
389. Horecker BL, Smyrniotis PZ (1953) J Am Chem Soc 75: 1009
390. Datta A, Racker E (1961) J Biol Chem 236: 624
391. Racker E, de la Haba G, Leder IG (1953) J Am Chem Soc 75: 1010
392. Racker E (1961) Transketolase. In: Boyer P, Lardy H, Myrbäck K (eds) The Enzymes. Academic Press, New York, vol. 5, p 397
393. Villafranca JJ, Axelrod B (1971) J Biol Chem 246: 3126
394. de la Haba G, Leder IG, Racker E (1955) J Biol Chem 214: 409
395. Sprenger GA, Schörken U, Sprenger G, Sahm H (1995) Eur J Biochem 230: 525
396. French C, Ward JM (1995) Biotechnol Lett 17: 247
397. Kotchetov GA (1982) Methods Enzymol 90: 209
398. Iida A, Teshiba S, Mizobuchi K (1993) J Bacteriol 175: 5375
399. Sprenger GA (1993) Biochim Biophys Acta 1216: 307
400. Chen J-H, Gibson JL, McCue LA, Tabita FR (1991) J Biol Chem 266: 20447
401. Fletcher TS, Kwee IL, Nakada T, Largman C, Martin BM (1992) Biochemistry 31: 1892
402. Sundstrom M, Lindqvist Y, Schneider G, Hellman U, Ronne H (1993) J Biol Chem 268: 24346
403. Metzger MH, Hollenberg CP (1994) Appl Microbiol Biotechnol 42: 319
404. McCool BA, Plonk SG, Martin PR, Singleton CK (1993) J Biol Chem 268: 1397
405. Lindqvist Y, Schneider G, Ermler U, Sundström M (1992) EMBO J 11: 2373
406. Nikkola M, Lindqvist Y, Schneider G (1994) J Mol Biol 238: 387
407. König S, Schellenberger A, Neef H, Schneider G (1994) J Biol Chem 269: 10879
408. Nilsson U, Lindqvist Y, Kluger R, Schneider G (1993) FEBS Lett 326: 145
409. Bolte J, Demuynck C, Samaki H (1987) Tetrahedron Lett 28: 5525
410. Dickens F, Williamson DH (1958) Biochem J 68: 74
411. Dickens F (1962) Biochem Prep 9: 86
412. Zimmermann F, Fessner W-D, unpublished
413. Dalmas V, Demuynck C (1993) Tetrahedron Asymmetry 4: 2383
414. Wood T (1973) Prep Biochem 3: 509
415. Mocali A, Aldinucci D, Paoletti F (1985) Carbohydr Res 143: 288
416. Effenberger F, Null V, Ziegler T (1992) Tetrahedron Lett 33: 5157
417. Kobori Y, Myles DC, Whitesides GM (1992) J Org Chem 57: 5899
418. Demuynck C, Bolte J, Hecquet L, Dalmas V (1991) Tetrahedron Lett 32: 5085
419. Ziegler T, Straub A, Effenberger F (1988) Angew Chem Int Ed Engl 27: 716
420. Effenberger F, Straub A, Null V (1992) Liebigs Ann Chem 1297
421. Hecquet L, Bolte J, Demuynck C (1994) Tetrahedron 50: 8677

422. Dalmas V, Demuynck C (1993) Tetrahedron Asymmetry 4: 1169
423. Csuk R, Glänzer BI (1991) Chem Rev 91: 49
424. Servi S (1990) Synthesis 1
425. Crout DHG, Dalton H, Hutchinson DW, Miyagoshi M (1991) J Chem Soc Perkin Trans I 1329
426. Bringer-Meyer S, Schmiz K-L, Sahm S (1986) Arch Microbiol 146: 105
427. Kuo DJ, Dikdan G, Jordan F (1986) J Biol Chem 261: 3316
428. Farrenkopf B, Jordan F (1992) Protein Express Purif 3: 101
429. Zehender H, Trescher D, Ullrich J (1987) Eur J Biochem 167: 149
430. Mücke U, König S, Hübner G (1995) Biol Chem Hoppe Seyler 376: 111
431. Neale AD, Scopes RK, Wettenhall REH, Hoogenraad NJ (1987) J Bacteriol 169: 1024
432. Kellermann E, Seeboth PG, Hollenberg CP (1986) Nucleic Acids Res 14: 8963
433. Hohmann S (1991) J Bacteriol 173: 7963
434. Hohmann S, Cederberg H (1990) Eur J Biochem 188: 615
435. Bräu B, Sahm H (1986) Arch Microbiol 144: 296
436. Conway T, Osman YA, Konnan JI, Hoffmann EM, Ingram LO (1987) J Bacteriol 169: 949
437. Neale AD, Scopes RK, Wettenhall REH, Hoogenraad NJ (1987) Nucleic Acids Res 15: 1753
438. Lowe SE, Zeikus JG (1992) J Gen Microbiol 138: 803
439. Alvarez ME, Rosa AL, Temporini ED, Wolstenholme A, Panzetta G, Patrito L, Maccioni HJF (1993) Gene 130: 253
440. Kelley PM, Godfrey K, Lal SK, Alleman M (1991) Plant Mol Biol 17: 1259
441. Dyda F, Furey W, Swaminathan S, Sax M, Farrenkopf B, Jordan F (1993) Biochemistry 32: 6165
442. Muller YA, Lindqvist Y, Furey W, Schulz GE, Jordan F, Schneider G (1993) Structure 1: 95
443. Abraham W-R, Stumpf B (1987) Z Naturforsch 42 C: 559
444. Chen GC, Jordan F (1984) Biochemistry 23: 3576
445. Bornemann S, Crout DHG, Dalton H, Hutchinson DW, Dean G, Thomson N, Turner MM (1993) J Chem Soc Perkin Trans I 309
446. Crout DHG, Littlechild J, Morrey SM (1986) J Chem Soc Perkin Trans I 105
447. Kren V, Crout DHG, Dalton H, Hutchinson DW, Konig W, Turner MM, Dean G, Thomson N (1993) J Chem Soc Chem Commun 341
448. Grue-Søresen G, Spencer ID (1988) J Am Chem Soc 110: 3714
449. Fuganti C, Grasselli P (1977) Chem Ind 983
450. Fuganti C, Grasselli P, Poli G, Servi S, Zorzella A (1988) J Chem Soc Chem Commun 1619
451. Karasek MA, Greenberg DM (1957) J Biol Chem 227: 191
452. Malkin LI, Greenberg DM (1964) Biochim Biophys Acta 85: 117
453. Greenberg DM (1967) Methods Enzymol 931
454. Bruns FH, Fiedler L (1958) Biochem Z 330: 324
455. Bruns FH, Fiedler L (1958) Nature 181: 1533
456. Naoi M, Takahashi T, Kuno N, Nagatsu T (1987) Biochem Biophys Res Commun 143: 482
457. Dainty RH (1970) Biochem J 117: 585
458. Vassilev VP, Uchiyama T, Kajimoto T, Wong C-H (1995) Tetrahedron Lett 36: 4081
459. Herbert RB, Wilkinson B, Ellames GJ (1994) Can J Chem 72: 114
460. Herbert RB, Wilkinson B, Ellames GJ, Kunec EK (1993) J Chem Soc Chem Commun 205
461. Ura D, Hashimukai T, Matsumoto T, Fukuhara N (1991) CA 115: 90657q
462. Anderson DM, Hsiao H-H (1992) Enzymatic method for L-serine production. In: Rozzell D, Wagner F (eds) Biocatalytic Production of Amino Acids & Derivatives. Hanser, Munich, p 23
463. Ulevitch RJ, Kallen RG (1977) Biochemistry 16: 5342
464. Schirch L, Peterson D (1980) J Biol Chem 255: 7801
465. Sakamoto M, Masuda T, Yanagimoto Y, Nakano Y, Kitaoka S (1991) Agric Biol Chem 55: 2243
466. Miyazaki SS, Toki S, Izumi Y, Yamada H (1987) Eur J Biochem 162: 533
467. Miyazaki SS, Toki S, Izumi Y, Yamada H (1987) Agric Biol Chem 51: 2587
468. Ide H, Hamaguchi K, Kobata S, Murakami A, Kimura Y, Makino K, Kamada M, Miyamoto S, Nagaya T, Kamogawa K, Izumi Y (1992) J Chromatogr 596: 203
469. Turner SR, Ireland R, Morgan C, Rawsthorne S (1992) J Biol Chem 267: 13528
470. Garrow TA, Brenner AA, Whitehead VM, Chen X-N, Duncan RG, Korenberg JR, Shane B (1993) J Biol Chem 268: 11910
471. Byrne PC, Sanders PG, Snell K (1992) Biochem J 286: 117

472. Jagath-Reddy J, Ganesan K, Saithri HS, Datta A, Rao NA (1995) Eur J Biochem 230: 533
473. Martini F, Angelaccio S, Pascarella S, Barra D, Bossa F, Schirch V (1987) J Biol Chem 262: 5499
474. Martini F, Maras B, Tanci P, Angelaccio S, Pascarella S, Barra D, Bossa F, Schirch V (1989) J Biol Chem 264: 8509
475. McNeil JB, McIntosh EM, Taylor BV, Zhang FR, Tang S, Bognar AL (1994) J Biol Chem 269: 9155
476. Chistoserdova LV, Lidstrom ME (1994) J Bacteriol 176: 6759
477. Urbanowski ML, Plamann MD, Stauffer LT, Stauffer GV (1984) Gene 27: 47
478. Chan VL, Bingham HL (1990) Gene 101: 51
479. McClung CR, Davis CR, Page KM, Denome SA (1992) Mol Cell Biol 12: 1412
480. Miyata A, Yoshida T, Yamaguchi K, Yokoyama C, Tanabe T, Toh H, Mitsunaga T, Izumi Y (1993) Eur J Biochem 212: 745
481. Plamann MD, Stauffer LT, Urbanowski ML, Stauffer GV (1983) Nucleic Acids Res 11: 2065
482. Rossbach S, Hennecke H (1991) Mol Microbiol 5: 39
483. Lotz BT, Gasparski CM, Peterson K, Miller MJ (1990) J Chem Soc Chem Commun 16: 1107
484. Saeed A, Young DW (1992) Tetrahedron 48: 2507
485. Chen MS, Schirch L (1973) J Biol Chem 248: 7979
486. Billhardt UM, Stein P, Whitesides GM (1989) Bioorg Chem 17: 1
487. Patel SS, Conlon HD, Walt DR (1986) J Org Chem 51: 2842
488. Ouyang T, Walt DR, Patel SS (1990) Bioorg Chem 18: 131
489. Patel SS, Walt DR (1987) J Biol Chem 262: 7132
490. Kumari S, Tishel R, Eisenbach M, Wolfe AJ (1995) J Bacteriol 177: 2878
491. Martin DP, Bibart RT, Drueckhammer DG (1994) J Am Chem Soc 116: 4660
492. Kao CM, Katz L, Khosla C (1994) Science 265: 509
493. Kao CM, Luo GL, Katz L, Cane DE, Khosla C (1994) J Am Chem Soc 116: 11612
494. McDaniel R, Ebert-Khosla S, Hopwood DA, Khosla C (1993) Science 262: 1546
495. McDaniel R, Ebert-Khosla S, Hopwood DA, Khosla C (1994) J Am Chem Soc 116: 10855
496. McDaniel R, Ebert-Khosla S, Hopwood DA, Khosla C (1995) Nature 375: 549
497. Tsoi CJ, Khosla C (1995) Chem Biol 2: 355
498. Shen B, Hutchinson CR (1993) Science 262: 1535
499. Rohr J (1995) Angew Chem Int Ed Engl 34: 881
500. Decker H, Haag S, Udvarnoki G, Rohr J (1995) Angew Chem Int Ed Engl 34: 1107
501. Effenberger F (1994) Angew Chem Int Ed Engl 33: 1555
502. Jorns MS (1979) J Biol Chem 254: 12145
503. Hochuli E (1983) Helv Chim Acta 66: 489
504. Becker W, Pfeil E (1966) Biochem Z 346: 301
505. Smitskamp-Wilms E, Brussee J, van der Gen A (1991) Recl Trav Chim Pays-Bas 110: 209
506. Lauble H, Muller K, Schindelin H, Forster S, Effenberger F (1994) Proteins 19: 343
507. Cheng IP, Poulton JE (1993) Plant Cell Physiol 34: 1139
508. Woker R, Champluvier B, Kula MR (1992) J Chromatogr 584: 85
509. Klempier N, Griengl H, Hayn M (1993) Tetrahedron Lett 34: 4769
510. van Scharrenburg GJM, Sloothaak JB, Kruse CG, Smitskamp-Wilms E, Brussee J (1993) Ind J Chem B 32: 16
511. Wajant H, Mundry KW (1993) Plant Sci 89: 127
512. Wajant H, Mundry KW, Pfizenmaier K (1994) Plant Mol Biol 26: 735
513. Xu L-L, Singh BK, Conn EE (1988) Arch Biochem Biophys 263: 256
514. Albrecht J, Jansen I, Kula MR (1993) Biotechnol Appl Biochem 17: 191
515. Hughes J, Decarvalho JPC, Hughes MA (1994) Arch Biochem Biophys 311: 496
516. Brussee J, Roos EC, van der Gen A (1988) Tetrahedron Lett 29: 4485
517. Becker W, Pfeil E (1966) J Am Chem Soc 88: 4299
518. Becker W, Freund H, Pfeil E (1965) Angew Chem Int Ed Engl 4: 1079
519. Niedermeyer U, Kula M-R (1990) Angew Chem Int Ed Engl 29: 386
520. Effenberger F, Ziegler T, Förster S (1987) Angew Chem Int Ed Engl 26: 458
521. Ognyanov VI, Datcheva VK, Kyler KS (1991) J Am Chem Soc 113: 6992
522. Brussee J, Loos WT, Kruse CG, van der Gen A (1990) Tetrahedron 46: 979
523. Huuhtanen TT, Kanerva LT (1992) Tetrahedron Asymmetry 3: 1223
524. Zandbergen P, van der Linden J, Brussee J, van der Gen A (1991) Synth Commun 21: 1387
525. Kiljunen E, Kanerva LT (1994) Tetrahedron Asymmetry 5: 311
526. Effenberger F, Hörsch B, Förster S, Ziegler T (1990) Tetrahedron Lett 31: 1249

527. Klempier N, Pichler U, Griengl H (1995) Tetrahedron Asymmetry 6: 845
528. Ziegler T, Hörsch B, Effenberger F (1990) Synthesis 575–578
529. Menendez E, Brieva R, Rebolledo F, Gotor V (1995) J Chem Soc Chem Commun 989–990
530. Effenberger F, Hörsch B, Weingart F, Ziegler T, Kühner S (1991) Tetrahedron Lett 32: 2605
531. Effenberger F, Eichhorn J, Roos J (1995) Tetrahedron Asymmetry 6: 271
532. Henderson I, Sharpless KB, Wong C-H (1994) J Am Chem Soc 116: 558
533. Badía J, Baldomà L, Aguilar J, Boronat A (1989) FEMS Microbiol Lett 65: 253
534. Card PJ, Hitz WD, Ripp KG (1986) J Am Chem Soc 108: 158
535. Berger A, de Raadt A, Gradnig G, Grasser M, Loew H, Stuetz AE (1992) Tetrahedron Lett 33: 7125
536. Look GC, Fotsch CH, Wong C-H (1993) Acc Chem Res 26: 182
537. Wong C-H, Halcomb RL, Ichikawa Y, Kajimoto T (1995) Angew Chem Int Ed Engl 34: 412
538. Pederson RL, Kim MJ, Wong C-H (1988) Tetrahedron Lett 29: 4645
539. Liu KKC, Kajimoto T, Chen L, Zhong Z, Ichikawa Y, Wong C-H (1991) J Org Chem 56: 6280
540. Zhou PZ, Salleh HM, Chan PCM, Lajoie G, Honek JF, Nambiar PTC, Ward OP (1993) Carbohydr Res 239: 155
541. Lees WJ, Whitesides GM (1992) Bioorg Chem 20: 173
542. Hung RR, Straub JA, Whitesides GM (1991) J Org Chem 56: 3849
543. Winchester B, Fleet GWJ (1992) Glycobiology 2: 199
544. Henderson I, Laslo K, Wong C-H (1994) Tetrahedron Lett 35: 359
545. Holt KE, Leeper FJ, Handa S (1994) J Chem Soc Perkin Trans I 231–234
546. Chou WC, Chen LH, Fang JM, Wong C-H (1994) J Am Chem Soc 116: 6191
547. Kuan KT, Weber DS, Sottile L, Goux WJ (1992) Carbohydr Res 225: 123
548. Fleischmann RD, et al. (1995) Science 269: 496
549. Marshall E (1994) Science 266: 208
550. Sakowicz R, Gold M, Jones JB (1995) J Am Chem Soc 117: 2387
551. Wilks HM, Moreton KM, Halsall DJ, Hart KW, Sessions RD, Clarke AR, Holbrook JJ (1992) Biochemistry 31: 7802
552. Wang QP, Graham RW, Trimbur D, Warren RAJ, Withers SG (1994) J Am Chem Soc 116: 11594
553. Oshima T (1994) Curr Opin Struct Biol 4: 623
554. Bocanegra JA, Scrutton NS, Perham RN (1993) Biochemistry 32: 2737
555. Mittl PRE, Berry A, Scrutton NS, Perham RN, Schulz GE (1994) Protein Sci 3: 1504
556. Reymond JL, Chen YW (1995) Tetrahedron Lett 36: 2575